scientific periodicals

their historical development,
characteristics and control

MECHANIC'S MAGAZINE.

VOL. 1.

OURS AND FOR US

KNOWLEDGE IS POWER.

scientific periodicals

their historical development, characteristics and control

by

BERNARD HOUGHTON MA FLA

Senior Lecturer in Information Work
LIVERPOOL POLYTECHNIC DEPARTMENT OF
LIBRARY AND INFORMATION STUDIES

LINNET BOOKS & CLIVE BINGLEY

Library of Congress Cataloging in Publication Data

Houghton, Bernard, 1935-
　　Scientific periodicals.
　　Includes bibliographical references and index.
　　1. Science—Periodicals. 2. Technology—Periodicals.
3. Scientific literature—Periodicals. 4. Technical
literature—Periodicals. I. Title.
Q1.H57　　　070.4'49'5　　　74-30193
ISBN 0-208-01363-6

FIRST PUBLISHED 1975
THIS EDITION SIMULTANEOUSLY PUBLISHED IN THE USA
BY LINNET BOOKS AN IMPRINT OF SHOE STRING PRESS INC
995 SHERMAN AVENUE HAMDEN CONNECTICUT 06514
PRINTED IN GREAT BRITAIN
COPYRIGHT © BERNARD HOUGHTON 1975
ALL RIGHTS RESERVED

Contents

		page
Chapter 1	The beginnings of the scientific journal	11
Chapter 2	The development and specialisation of the scientific journal	20
Chapter 3	Present forms of the scientific and technical journal	32
Chapter 4	The scientific journal: some problems and the alternatives	42
Chapter 5	Access to the journal literature—1: catalogues, bibliographies and lists of journals	52
Chapter 6	Access to the journal literature—2: abstracts—their nature and use	68
Chapter 7	Access to the journal literature—3: abstracting services—their development, some problems and solutions, and the impact of the computer	81
Chapter 8	Characteristics of the literature: growth, obsolescence, citations patterns and scattering	101
Index		129

Definition

The term periodical is currently defined as a type of serial in which the parts (called issues) are generally characterised by variety of contents and contributors, both within the issue and from one issue to another. With the general exception of newspapers and some other types of popular periodical, issues are commonly designed and numbered as constituents of a notional volume, which is completed at determined intervals by the issue of a volume title-page and/or index. Commonly used synonymous terms for periodical are journal and magazine.

List of illustrations

page

Frontispiece Title-page of *Mechanic's magazine,* vol one

Figure 1 Title-page and first page of text of *Journal des scavans,* vol one   13

2 Title-page of *Philosophical transactions of the Royal Society,* vol one   15

3 Volume one, number one of the *Mechanic's magazine*   22

4 Page one of *Nature,* vol one   25

5 Contents list of first issue of *Nature*   26

6 Title-page of *Minutes of proceedings of the Institution of civil engineers,* vol one   28

7 Contents list of *Minutes of proceedings of the Institution of civil engineers,* vol one   29

8 Examples of indicative abstracts from *Engineering index*   70

9 Examples of informative abstracts from *Biological abstracts*   71

10 Overlap in the lists of journals monitored by *Biosciences Information Service* (BISOS), *Chemical Abstract Service* (CAS) and *Engineering index* (EI)   87

11 Expansion in the numbers of papers abstracted by *Chemical abstracts, Biological abstracts, Engineering index* and *Physics abstracts*   92

12 An example of an INSPEC SDI profile   94

		page
13	Output references for the profile illustrated in figure 12	95
14	An example of KWIC indexing taken from *Chemical titles*	98
15	Graph of cumulative numbers of journals published from 1665 to 1970	101
16	Obsolescence graph for the journal literature of physics	108
17	How to use the *Science citation index* (SCI)	112
18	How to use the *Permuterm index* (SCI)	114
19	A one-step journal citation map	116
20	A two-step journal citation map	117
21	A paper citation network	120
22	Graph of the cumulative distribution of papers over journals	123

Introduction

This work is offered as an introduction to the historical development, forms, bibliographical control, and structure and characteristics of the periodical literature of science and technology.

It is hoped that the text will be of value to students of scientific librarianship and information work, to practising information scientists, and to those members of the scientific community who wish to obtain a conspectus of the major primary and secondary forms of the literature of science and technology—the periodical literature.

The sections on the evolution of the periodical attempt to identify the major landmarks, trends and problems in the development of the literature, as do those on secondary sources and bibliographical control. The section on the characteristics of the literature is not intended to be an exhaustive survey of growth, obsolescence and citation patterns, etc, but is given as a review of some of the significant primary source papers on these topics.

My thanks are due to the staffs of Liverpool Technical Library, Liverpool Polytechnic Library, the British Library Lending Division, the Science Reference Library, and the Aslib library for their invaluable assistance in making available the documents needed during the compilation and writing of this book. 1 would also like to acknowledge the assistance given by the librarians of the Institution of Civil Engineers and the Royal Society of Great Britain in providing photographic copies of the pages of their journals reproduced as illustrations in this work, and to the Biosciences Information Service, Engineering Index Inc, INSPEC, Science Citation Index, and the United Kingdom Chemical Information Service for granting me permission to reproduce examples of pages from their publications.

Finally I would like to thank Glyn Rowland, my colleague in Liverpool Polytechnic Department of Library and Information Studies, and Mrs Carolyn Prosser, who recently worked in the department as my research assistant for their constructive critical comments which I often called upon when writing the text.

Liverpool
September 1974

BERNARD HOUGHTON

I

The beginnings of the scientific journal

THE ORIGINS of the modern scientific periodical lie in the development of the newspaper and the establishment of the scientific society. It has been reported that a Chinese newspaper, the *Peking gazette*, was founded in the seventh century AD but the antecedents of the newspaper in the western world can be traced to the Roman *Acta diurna* which were posted daily in public places to give the populace information on government decrees, news of battles and athletic contests, etc. Fielding H Garrison maintains that the earliest identifiable newspaper, as distinct from the ephemeral newsbooks and broadsheets, was the *Mercurius Gallo-belgicus* which commenced publication in Cologne in 1594 and covered European news from 1588 to 1594.[1] This was imitated on the continent by several 'mercurial type' publications. The name *Mercury* became current in England to denote the newsbook or coranto publications issued from time to time by publishers to disseminate information on single events of popular interest. It was out of these publications that the English newspaper evolved. Other precursors of the newspaper in England were the 'newsletters' issued in the fourteenth and fifteenth centuries by the Scriveners Company of London to supply people of rank with daily news of current events. It has been claimed that the first English newspaper to be published was the *Corante, or newes from Italy, Germany, Hungarie, Spain and France*, six issues of which were published in London with a weekly frequency during September and October 1621. Another notable example of the early newspaper was the *Frankfurter journal*, published from 1615.

The establishment and development of the scientific society was the other factor instrumental in the emergence of the journal as the main medium of scientific communication. Garrison observes that the 'vague seventeenth century appetite for new knowledge manifesting itself as the curiosity about sights, portents, marvels, monstrosities and freaks of nature transformed itself into a widespread intellectual uplift aiming at no less than complete control of the best knowledge available'.[1] The scientific society was nurtured in this enlightened atmosphere. Francis Bacon had in the sixteenth century advocated the cooperative action of natural philosophers, as scientists were then termed. He purported that philosophers should 'read not to contradict and confute, nor to believe and take for granted ... but to weigh and consider'. During the first fifty years of the seventeenth century the 'hidden colleges', informal networks of philosophers, were developed through personal contact and private written communication. These colleges were to become the formalised academies and societies which started to record and disseminate knowledge of the experiments of their members in minute books and through written communication between members. Inevitably a pattern of communication emerged in which one experiment equalled one communication although the well-structured scientific paper as it is known today did not appear until after the establishment of the specialised scientific journals between 1780 and 1790. The practice of communication by letter was restricting in that the information could be disclosed to only one person at a time or to a very limited number if copyists were employed. Publication in conventional book form was cumbersome, time consuming and uneconomic as the author was forced to delay until he had gathered enough matter to justify publication.

The *Journal des scavans* is generally cited as the first scientific journal. This was founded by de Sallo, a counsellor of the French court of parliament. The Marquis de Crenan, an associate of Pascal, the mathematician and philosopher, made the following comment in a letter to Huygens. 'M de Sallo ... desires to have correspondence through Europe of all events as much in affairs of state as in matters of science. He has asked me to write to you requesting your approval for an exchange of news to this end; he is a person of merit and consideration.'[2] De Sallo, a man of prodigious energy, kept two scriveners continuously at work transcribing the most significant passages he encountered during his extensive readings. He epitomised the new learning and waged a constant battle against the superstitions and prejudices rife at the time. The abstracts and notes made by his scriveners were ultimately incorporated into the *Journal des scavans*, the first issue of which appeared on 5th January 1665. This issue consisted of twenty pages, including ten articles, some letters and notes. The avowed purpose of the journal was 'to catalogue and give useful information on books published in Europe and to summarise their works, to make known experiments in physics, chemistry and anatomy that may serve to explain natural

LE
JOURNAL
DES
SCAVANS,

De l'An M. DC. LXV.

Par le Sieur

DE HEDOUVILLE.

A AMSTERDAM,
Chez PIERRE LE GRAND.
M. DC. LXXXIV.

I.
JOURNAL
DES SÇAVANS,

Du Lundy 5 Janvier, M. DC. LXV,
Par le Sieur DE HEDOUVILLE.

Victoris Vitensis, & Vigilii Tapsensis, Provinciæ Bisacenæ Episcoporum Opera; Edente R. P. Chifletio, Soc. Jesu Presb. in 4. Divione.

E seul ouvrage qui nous reste de Victor Vitensis est l'histoire de la persecution d'Afrique sous les Wandales. On voit par le commencement de cette histoire qu'il l'escrivit l'an 487. Nous avions deja cet ouvrage dans la Bibliotheque des Peres, sous le nom de Victor Vitensis : mais tous les sçavans demeurent presentement d'accord, qu'il est de Victor Vitensis. De plus, cette histoire estoit defectueuse dans la Bibliotheque des Peres : car on n'y voit point la liste des Evéques d'Afrique qui se trouverent enveloppez dans cette persecution. Cependant c'est une piece excellente, & qui peut beaucoup servir à l'éclaircissement de plusieurs difficultez de l'histoire Ecclesiastique. C'est pourquoi cette 1665. A 5 édi-

FIGURE 1: *Title-page and first page of text of* Journal des scavans, *vol one.*

phenomena, to describe useful or curious inventions or machines and to record meteorological data, to cite the principal decisions of civil and religious courts and censures of universities, to transmit to readers all current events worthy of the curiosity of men'. De Sallo claimed that his journal was instituted 'for the relief of those either too indolent or too occupied to read whole books', he saw the publication as 'a means of satisfying curiosity and becoming learned with little effort'. The popularity of the journal attracted the attention of the government and it was for a little while suppressed for printing material which offended the crown. Its royal privilege permitting publication was revoked for a period in the first year of publication. The journal was however relicensed the following year and publication with varying frequencies continued under the original title until 1816 when it became the *Journal des savants,* still a leading periodical, but now of a literary nature. The initial success of *Journal des scavans* can be gauged from the fact that reprint editions appeared in Holland between 1665 and 1792 and in Germany between 1667 and 1671.

The first English scientific journal was published only three months after the appearance of the French journal. An eminent group of English philosophers, including Boyle, Hooke, Moray and Oldenburg, then secretary of the Royal Society of London, saw the need for a truly scientific journal which would, unlike the *Journal des scavans,* exclude legal and theological questions but which would be used to record experiments conducted by members of the Royal Society and to publish selections from their correspondence with their counterparts in Europe. The Council of the Royal Society decreed that the *Philosophical transactions* should be printed on the first Monday of each month, 'if it has sufficient matter for it'. The first issue of the *Philosophical transactions* appeared on 6th May 1665 and consisted of sixteen pages comprising a dedication to the society, nine articles, a selective listing of current philosophical books and extracts from Oldenburg's voluminous foreign correspondences. Initially the direction, composition and publication of the transactions were the private responsibility of Oldenburg. It was not until the

PHILOSOPHICAL TRANSACTIONS:

GIVING SOME ACCOMPT

OF THE PRESENT Undertakings, Studies, and Labours

OF THE

INGENIOUS

IN MANY CONSIDERABLE PARTS OF THE

WORLD.

Vol I.
For *Anno* 1665, and 1666.

In the *SAVOY*,
Printed by *T. N.* for *John Martyn* at the Bell, a little without *Temple-Bar*, and *James Allestry* in *Duck-Lane*, Printers to the *Royal Society*.

Presented by the Author May. 30th 1667.

FIGURE 2: *Title-page of* Philosophical transactions of the Royal Society, *vol one.*

publication of the forty seventh volume that the transactions became the official organ of the Royal Society. The interests of most of the early contributors to the publication ranged through varied fields. An outstanding example of this span of interests is Robert Hooke, the founder of scientific metrology, whose published papers cover astronomy, geology, architecture and microscopy. The fragmentation of science and the rigid compartmentalisation of scientists into particular disciplines was not to occur until well into the next century. The *Philosophical transactions of the Royal Society* has an unbroken history except for a brief period of dormancy between 1676 and 1683. It became a model for many subsequent learned society journals and a standard for publications recording the results of scientific enquiry.

Two of the early German scientific journals, *Miscellanea curiosa*, 1670-1705 and *Acta eruditorum*, 1682-1731, were mirrors of the English and French originals. The *Miscellanea*, which is generally cited as the first German scientific journal, was published by the Collegium Natural Curiosum, the oldest society, with a continuous history dating from 1652. The *Miscellanea* printed contributions relating chiefly to medicine but also included botanical, mineralogical and zoological material. Contributions were encouraged from all parts of Europe and this outlooking policy can be seen in an analysis of its contributors. In the first ten years of the journal's existence ninety eight of its contributors were members of the Collegium while one hundred and ninety eight were non-members. The *Acta eruditorum* was a less scientific journal than the *Miscellanea* with only about one third of its content dealing with what would today be called science.

De Sallo's *Journal des scavans* become the model for publications which sought to appeal to a wide readership with broad interests extending into law and theology. The *Receuil des memoires et conferences sur les arts et les sciences* was published between 1672 and 1673 in the manner of and as a supplement to *Journal des scavans* and consisted largely of reviews of books, some original contributions, letters and extracts from other journals, notably *Philosophical transactions of the Royal Society.* The *Giornale de litterati d' Italia* commenced publication in 1668 and was another imitation of de Sallo's journal.

The *Acta medica et philosophia hafniensia* appeared in Copenhagen from 1673 to 1680 and like the *Miscellanea curiosa* was strongly medical in character. This journal served as a forum for disseminating the ideas of Thomas Bertholm, a celebrated physician who was professor of anatomy at the University of Copenhagen. Bertholm can be considered as a precursor of the critical editors of the nineteenth century in that he often embellished the communications he published in the *Acta* with his own comments and notes. A feature of almost all of the early scientific journals was their lack of critical comment. This evolved only slowly with the advent of the specialised chemical journals in the nineteenth century.

Many of the journals which appeared in the seventeenth and early eighteenth centuries floundered after one or two years and in many cases after only one or two issues. A significant reason for their early demise is that they were not underpinned by a sponsoring society: the scientific societies were founded mainly from the middle of the eighteenth century. Prior to this time the existing societies were content to publish their accounts in journals which were usually managed and financed by individuals, or to issue their works in the form of retrospective collections of papers at long and irregular intervals. The Academie des Sciences in Paris did not publish its regular proceedings, the *Histoires et memoires,* a journal which did much to reinforce a tradition of scientific publication established by the Royal Society's *Philosophical transactions,* until 1702. Before this date the reports of the work of its members appeared in the *Journal des scavans.* In 1750 the Academie founded a new series to publish the works of scientists who were not members of the society, the *Memories de mathematiques et de physique presentees a l'academie royale des sciences par divers scavans et lus dans ses assemblees.*

In addition to the lack of sponsoring societies other constricting factors on the growth of the journal literature were inherent in the social conditions of the time. Material was difficult to obtain because of problems in communications and there was a limit to the scientific output of the seventeenth and eighteenth centuries while a tradition of scientific research was being established.

The other significant early journals were the *Raccolta d'opuscoli scientifici e fililogici*, published in Venice from 1728 to 1757 and continued as the *Nuova raccolta opuscoli scientifici* from 1755 to 1787, *Le pour et le contre*, published in Paris from 1733 to 1740 and the *Gottingische zeitung von gelehrten sachen* issued from 1739 to 1752 but continued under varying titles well into the nineteenth century. A journal which chose to publish only original contributions in the field of natural science was *Der naturforscher* published at Halle from 1744 by Johann Ernst Immanuel Walch (1725-1778) and continued after the editor's death until 1804. The Abbe Rozier's *Observations* . . . again sought to publish only original material. This journal flourished between 1773 and 1794 and became probably the most important medium for the exchange of scientific ideas in the last quarter of the eighteenth century—its influence was seminal. The *Conferences* were designed as a repository of science rather than an organ of education: a conception which was characteristic of very few of the then existing journals. Indeed few of the early scientific journals contained only papers communicating the results of original experiments. Many of them were of the digest type of publication which claimed 'read us and it is not necessary to read the others because we give you the best'. Even in the eighteenth century scientists were bemoaning the confusion caused by the proliferation of digest journals—the so-called 'information explosion' has a long history. Rozier stated in his journal that he would not publish mere 'undigested compilations' or contributions which were 'merely agreeable' or offered to amuse the 'idle amateur'.

By the end of the seventeenth century about thirty scientific and medical journals had been published. This number was to expand rapidly throughout the following century. Garrison's listing of the scientific and medical periodicals published up to 1800 identifies seven hundred and one individual items although some of the titles he includes are hardly periodicals in the now accepted sense, many of them being reprinted editions of the minute books of societies and digests issued in parts. Garrison's list shows the preponderance of German titles which grew out of the German traits of thoroughness, diligence and method. This German monopoly of early scientific experimentation can also be seen in the later development of the specialised literature of physics and chemistry. Garrison estimates that out of 755 titles which appeared up to the close of the eighteenth century 401 were German, 96 French, 50 English, 43 Dutch and 37 Swiss.[1]

The journal had become the accepted medium of scientific communication by the middle of the eighteenth century and its functions were becoming clearly identifiable. The fragmentation of the journal into disciplinary divisions was by this time beginning but the specialisation by function had yet to emerge. The roles of the journal were now established:

1 they provided the scientific community and interested layity with news in the vernacular of work previously reported in foreign languages;

2 they provided the means for scientific and literary men to discourse on scientific work without having read the complete original account;

3 they conserved material which would otherwise have been dispersed through publication in individual tracts or pamphlets;

4 they aided scholarship by providing inexpensive channels of communication;

5 they encouraged scientists to publicise their work;

6 they offered a forum for the continuous critical examination of scientific hypotheses and theories.

REFERENCES

1 Garrison, Fielding H 'The medical and scientific periodicals of the seventeenth and eighteenth centuries' *Bulletin of the history of medicine. Johns Hopkins University* 2 (5), July 1934, pp 285-341.

2 Porter, J R 'The scientific journal: 300th anniversary' *Bacteriological reviews* 28 (3), Sept 1964, p 217.

2

The development and specialisation of the scientific journal

AS WITH THE EARLY general scientific journals the first specialised chemical journals were published by individuals rather than under the auspices of learned societies. Lorenz von Crell, professor of chemistry at Helmstadt University, produced a series of specifically chemical journals, notably *Chemisches journal fur die freunde der naturlehre, arzneygelahrtheit, haushaltungskunst und manufacturen*, 1778-1781 and *Die neuesten entdeckungen in der chemie*, 1781-1786. The oldest chemical journal in continuous existence is *Annales de chemie*, which commenced publication in 1789 and was the journal which advocated the new chemistry of Lavoisier disproving the phlogistron theory. The *Annales* documented many of the important contributions to the development of chemistry over the subsequent fifty years. In 1815 it became *Annales de chemie et de physique* and in 1914 split into two publications, *Annales de chemie* and *Annales de physique*. Other important early chemical journals were the publications issued by Alexander Niklaus Scherer, particularly *Allgemeines journal der chemie*, 1798-1803, which published research papers in pure and applied chemistry. This publication remained in existence, with shifts in editorial policy and title, until 1834 when it became *Journal fur praktische chemie*, a publication still extant. Scherer's other important contribution was the *Archiv fur die theoretische chemie*, which served as a platform of discussion for adherents of the new ideas of Lavoisier. The first purely chemical journal to be published in England was *The chemist*, which appeared between 1824 and 1825 'to give an outline of the principles of chemistry . . . to make *The chemist* a repository of every valuable discovery, either in chemistry or the sciences connected with it which might be made either at home or abroad'.

The early development of organic chemistry was documented

largely in the publication edited by Justus Liebig, *Annalen der chemie*, which commenced in 1832. Liebig was a rigid authoritarian who felt it his duty to criticise the papers he published in his journal and he insisted that all ideas must be supported by empirical data and experimental detail. Indeed criticism is a mild word when applied to Liebig; any opponent was publicly crushed and lampooned in the *Annalen*.[1] The publications of the chemical societies followed in the wake of the private compilations. The Chemical Society issued its *Memoires and proceedings* from 1841 to 1847 when they became the *Journal of the Chemical Society*, the *Journal de la Societe Chimique de France* was issued from 1857 and the *Proceedings of the American Chemical Society* appeared in 1876 appropriately timed to publish the work of the army of American chemists then returning from the German technological universities.

The first specialised physical journal to be published was *Journal der physik* which was issued at Halle and Leipzig from 1790. In England the *Philosophical magazine*, a journal devoted to physics which is still extant and prestigious, was founded in 1798. Again both of these publications were independent of the learned societies; as in the general and chemical fields, the society publications appeared after the establishment of privately owned journals. The *Proceedings of the Royal Physical Society of Edinburgh* were instituted in 1854, the *Proceedings of the London Mathematical Society* in 1865 and the *Proceedings of the Physical Society of London* in 1874.

Throughout the eighteenth and nineteenth centuries the spread of specialised journals continued into the developing fields of science. The *Botanical magazine* was founded in 1746 and ultimately became *Curtis's botanical magazine*, the first illustrated botanical journal and a publication whose exquisitely designed colour plates were to influence illustration in journals relating to all fields of science and technology. This journal can in some measure be considered as a precursor of the modern illustrated technical journals. The *Annals of natural history, or magazine of zoology, botany and geology* was first published in 1838. In 1840 this publication merged with the *Magazine of*

Mechanic's Magazine,
Museum, Register, Journal, & Gazette.

——— Industry! rough power!
Whom labour still attends, and sweat, and pain;
Yet, the kind source of every gentle art,
And all the soft civility of life.
Thomson.

No. 1.]   SATURDAY, AUGUST 30, 1823.   [Price 3d.

JAMES WATT.

MEMOIR OF JAMES WATT,
THE GREAT IMPROVER OF THE STEAM-ENGINE.

Many great and distinguished men, the ornaments of the last and present centuries, have been more known and much more talked of than James Watt; but, perhaps, no one of them was the fortunate author of so much real good to mankind or has equal claims on their gratitude. Now, indeed, it is generally known, that he was one of the most successful and skilful inventors of machinery of the age. His good fortune may encourage, and his perseverance instruct the present and all future generations of mechanics; and, therefore, his biography has been selected, as it seems particularly well adapted, for the first number of a work which is to be entirely devoted to their amusement an improvement. Mr. Watt was also a kind good-hearted man—giving lustre to his art; not only by the prodigious power he created, but by the life he led. He acquired wealth and honour by his own exertions, and was praised for his wisdom as well as for his skill. Though we do not pretend to assert that there is any thing in mechanical pursuits which pecu-

B

FIGURE 3: *Vol one, number one of the* Mechanic's magazine.

natural history and journal of zoology, botany, mineralogy, geology and meteorology to become the *Annals and magazine of natural history*.

The trade and technical press evolved concurrently with the growth of the industrial revolution although before the nineteenth century there were isolated examples of publications such as *Collection for improvement of husbandry and trade* 1691, which anticipated the artisans' journals established a century later. One of the first general trade journals to be published in England was the *Mechanics' magazine,* founded in 1823 to give mechanics 'a better acquaintance with the history and principles of the arts they practice'. The magazine contained 'accounts of new discoveries, inventions and improvements, with illustrated drawings, explanations of secret processes, economical receipts, practical applications of mineralogy and chemistry, plans and suggestions for the abridgement of labour, reports on the state of the arts in this and other countries, memoires and occasionally portraits of eminent mechanics, etc'. The *Mechanics' magazine* was an offshoot of the Mechanics Institute movement and consequently its contents were directed at the artisan classes. This journal can be cited as the real precursor of today's technical and trade journals. The *American mechanics' magazine,* a frank imitation of the London publication, was founded in 1825 as 'a digest of mechanics and scientific progress'. This subsequently became the *Journal of the Franklin Institute,* a publication still extant, which was of particular importance in the early years of the nineteenth century as it included descriptions of newly granted American patents. Prior to 1943, the United States Patent Office omitted details of the claims of its patent grants from its official publications. Another important technical journal, the *Scientific American,* was published from 1845. This commenced its life as 'an advocate of industry and journal of mechanical and other improvements'. The *Scientific American* has survived and indeed flourished but its mechanical bias and original pioneering spirit which sought to debunk cranks and quacks have long since disappeared. The journal has for many years concentrated on publicising the importance and relevance of science and technology to the community in general. A similar mission has been pursued in the United Kingdom by the organs of the British Association for the Advancement of Science, its *Report of the British Association for the Advancement of Science* appeared from 1831 and this continued its life from 1938 as *Advancement of science*.

By the middle of the nineteenth century over one thousand scientific and technical journals were being published throughout the world. The *Engineer* was founded in 1856 not 'to furnish a dry register of progress of machinery' but 'to represent effectively the industrial activity in which we live, to keep pace with progress of improvements and developments in all departments of the arts and manufactures which contribute to our material comfort'. This journal was intended for the professional man rather than the artisan. Its first editorial considering the interdependence of science and technology can well be used today to illustrate to the specialist, whatever his field, the importance of maintaining the literature of *all* areas of science and technology. 'The professional engineer finds indeed that the data with which he works are to be gathered alike from the most homely percepts of every-day experience, and the remotest provinces of the physical sciences. He is necessarily a mechanic but not a mechanic merely; he has need of the deductions of chemistry as well as—as far at least as they are subservient to his precise knowledge—the material properties and agencies which it is his business to direct'. The other important early weekly publication, *Engineering*, which was to record the progress of the industrial revolution and to become with its rival compulsory reading for the engineer who wished to keep pace with progress, was published from 1866.

The general technical and scientific journals discussed above are largely those which survived, but the middle of the nineteenth century was littered with dead or dying journals which had attempted to capture the readership of a growing scientific community. Although the previous casualties were legion, Norman Lockyer, a brilliant young scientist who at the age of 33 had been elected a Fellow of the Royal Society, persuaded Macmillans to sponsor yet another new general scientific journal. His cause

A WEEKLY ILLUSTRATED JOURNAL OF SCIENCE

*"To the solid ground
Of Nature trusts the mind which builds for aye."*—WORDSWORTH

THURSDAY, NOVEMBER 4, 1869

NATURE: APHORISMS BY GOETHE

NATURE! We are surrounded and embraced by her: powerless to separate ourselves from her, and powerless to penetrate beyond her.

Without asking, or warning, she snatches us up into her circling dance, and whirls us on until we are tired, and drop from her arms.

She is ever shaping new forms: what is, has never yet been; what has been, comes not again. Everything is new, and yet nought but the old.

We live in her midst and know her not. She is incessantly speaking to us, but betrays not her secret. We constantly act upon her, and yet have no power over her.

The one thing she seems to aim at is Individuality; yet she cares nothing for individuals. She is always building up and destroying; but her workshop is inaccessible.

Her life is in her children; but where is the mother? She is the only artist: working-up the most uniform material into utter opposites; arriving, without a trace of effort, at perfection, at the most exact precision, though always veiled under a certain softness.

Each of her works has an essence of its own; each of her phenomena a special characterisation; and yet their diversity is in unity.

She performs a play; we know not whether she sees it herself, and yet she acts for us, the lookers-on.

Incessant life, development, and movement are in her, but she advances not. She changes for ever and ever, and rests not a moment. Quietude is inconceivable to her, and she has laid her curse upon rest. She is firm. Her steps are measured, her exceptions rare, her laws unchangeable.

She has always thought and always thinks; though not as a man, but as Nature. She broods over an all-comprehending idea, which no searching can find out.

Mankind dwell in her and she in them. With all men she plays a game for love, and rejoices the more they win. With many, her moves are so hidden, that the game is over before they know it.

That which is most unnatural is still Nature; the stupidest philistinism has a touch of her genius. Whoso cannot see her everywhere, sees her nowhere rightly.

She loves herself, and her innumerable eyes and affections are fixed upon herself. She has divided herself that she may be her own delight. She causes an endless succession of new capacities for enjoyment to spring up, that her insatiable sympathy may be assuaged.

She rejoices in illusion. Whoso destroys it in himself and others, him she punishes with the sternest tyranny. Whoso follows her in faith, him she takes as a child to her bosom.

Her children are numberless. To none is she altogether miserly; but she has her favourites, on whom she squanders much, and for whom she makes great sacrifices. Over greatness she spreads her shield.

She tosses her creatures out of nothingness, and tells them not whence they came, nor whither they go. It is their business to run, she knows the road. Her mechanism has few springs—but they never wear out, are always active and manifold.

The spectacle of Nature is always new, for she is always renewing the spectators. Life is her most exquisite invention; and death is her expert contrivance to get plenty of life.

She wraps man in darkness, and makes him for ever long for light. She creates him dependent upon the earth, dull and heavy; and yet is always shaking him until he attempts to soar above it.

B

FIGURE 4: *Page one of* Nature, *vol one.*

had the support of the scientific establishment of the day; Lyon Playfair, Benjamin Brodie, William Crookes, William Thompson and many others. The new weekly journal, *Nature*, first appeared on 4th November 1869. The stated objects were 'first to place

CONTENTS.

	PAGE
GOETHE: APHORISMS ON NATURE. By Prof. HUXLEY, F.R.S.	9
ON THE FERTILISATION OF WINTER-FLOWERING PLANTS. By A. W. BENNETT, F.L.S. (*With Illustrations*)	11
PROTOPLASM AT THE ANTIPODES	13
THE RECENT TOTAL ECLIPSE IN AMERICA. By J. NORMAN LOCKYER, F.R.S. (*With Illustrations*.)	14
MADSEN'S DANISH ANTIQUITIES. By SIR J. LUBBOCK, BART., F.R.S.	15
NEWMAN'S BRITISH MOTHS. By W. S. DALLAS, F.L.S (*With Illustrations*)	16
OUR BOOK SHELF	17
SCIENCE-TEACHING IN SCHOOLS. By the REV. W. TUCKWELL	18
THE LATE PROFESSOR GRAHAM. By Prof. WILLIAMSON, F.R.S. (*With Portrait*)	20
MEETING OF THE GERMAN NATURALISTS AND PHYSICIANS AT INNSBRUCK. By A GEIKIE, F.R.S.	22
TRIASSIC DINOSAURIA. By Prof. HUXLEY, F.R.S.	23
CORRESPONDENCE:—The Suez Canal. T. LOGIN, C.E.	24
NOTES	25
ASTRONOMY.—Astronomical Congress at Vienna	26
CHEMISTRY.—Abstracts of Papers by Bettendorff, Paterno, Peligot, &c.	27
PHYSICS.—Magnus on Heat Spectra	28
PHYSIOLOGY.—Pettenkofer on Cholera, &c.	28
SOCIETIES AND ACADEMIES	29
DIARY	30
BOOKS RECEIVED	30

FIGURE 5: *Contents list of the first issue of* Nature.

before the general public the results of scientific work and discovery and to urge the claims of science to a more general recognition in education and daily life; and secondly to aid scientific men themselves by giving early information of all advances made in any branch of natural knowledge throughout the world and by affording them an opportunity of discovering the various scientific questions which arise from time to time. Its original aims were realised and *Nature* has since become the most influential of all scientific journals. Its former editor John Maddox maintains that the journal survived its early years because of 'the great passion of its contributors for great causes which are still close to the heart of the scientific community' and also because 'Lockyer, the first editor, had a flair for journalism at its best'.² For over one hundred years *Nature* has maintained that 'discoveries are only discoveries when they are recognised as such outside the circle in which they are first made'.² Nevertheless *Nature's* character has changed throughout its life. Originally it was published as a journal 'popular in part, but also sound, and part devoted to scientific men and their intercourse'.³ After the first world war *Nature* slowly ceased to be a journal which sought to cater for the interested amateur and became the major international organ for announcing scientific discovery. Although almost miraculously it also remained the 'means of "placing before the general public the grand results of scientific work and scientific discovery" even if the going was at times a little hard'.² The journal is now truly international in flavour: two thirds of the copies now printed are purchased outside the United Kingdom and contributions and communications are received from universities and research centres in all parts of the world. After the second world war *Nature's* speed of publication became one of its distinguishing features and the quality which gave scientists the opportunity to be seen to be first in their fields to a world-wide audience.

The early scientific societies were catholic in their activities, and their publications, which included papers on all branches of science, engineering and philosophy, reflected their broad sweep of activity. The earliest of the specialised societies were in the main concerned with natural history. The Linnean Society was founded in 1788, for the 'cultivation of the science of natural history in all its branches and more especially of the natural history of Great Britain and Ireland'. The *Journal of the Linnean Society* did not appear until 1856. The Zoological Society of London was established in 1826 with its *Proceedings* dating from 1830, the Royal Entomological Society in 1833, with its *Transactions* being established in the following year, and the Botanical Society in 1836.

The oldest professional engineering institution in the world was founded by Thomas Telford, who with a group of his friends formed a society for engineers in England on 2nd January 1818 to provide a place where young civil engineers might become

MINUTES OF PROCEEDINGS

OF THE

INSTITUTION

OF

CIVIL ENGINEERS;

CONTAINING

ABSTRACTS OF THE PAPERS

AND OF THE

CONVERSATIONS,

FOR THE SESSION OF 1837.

———

LONDON:
PRINTED FOR THE INSTITUTION.
1837.

FIGURE 6: *Title-page of* Minutes of proceedings of the Institution of civil engineers, *vol one*.

ORIGINAL COMMUNICATIONS.

Baker, George. Description and model of a new Railway Chair.

Ballard, C. Description and model of a mode of framing Lock-gates.

Borthwick, M.A. Memoir on the use of Iron Piling, with drawings.

Bourns, C. Historical account of, and observations on the Port of London.

Bray, W. B. Drawings of the Pont des Invalides, the Pont de Jena, and the Pont du Carrousel, at Paris.

Bremner, J. Drawing and description of a machine used in forming the Harbour of Sarclet, Scotland, and drawing and description of a machine for clearing away sand from foundations, in harbour building.

Carnegie, Lindsey. Four drawings of Hunter's stone-planing machine.

Daglish, R. Drawing of a parallel rail and pedestals.

Hawkshaw, J. Paper on the mode of working Mines in South America.

Hopkins, Rice. Account of the Bodmin and Wadebridge Railroad.

Hopkins, T. Description and drawing of Ruthern Bridge, on the Bodmin and Wadebridge Railroad.

Marconi, E. Description and drawing of a proposed Floating Bridge, over the Vistula, to connect Warsaw and Prague.

Moreland, R. Drawing of an improved Mashing machine.

Morton, G. Description and drawing of a new Railway Carriage.

Page, Colonel. Papers on Canal navigation.

Perkins, Jacob. Three papers on the elasticity of Steam.

——————— On the causes of the difference of duty done by the Cornish and Soho engines, and on the improvement of Steam Boilers.

FIGURE 7: *Contents list of* Minutes of proceedings of the Institution of civil engineers, *vol one.*

acquainted with the art and practice of civil engineering. In the early years of the nineteenth century engineering was considered in England as merely a trade; there was a need for a body which would bestow professional status on the engineer. The original minute book records the foundation of 'the Institution of Civil Engineers for facilitating the acquirement of knowledge necessary in their profession and for promoting mechanical philosophy'. Engineering had not yet fragmented into its present numerous specialisations. Civil engineers were concerned with 'the art of directing the great sources of power in nature for the use and convenience of man' which included the construction of machinery, while military engineers devoted their energies to devising implements for waging war. The founding members were few and only limited finances were available, thus members were forced to publish their papers in the few current engineering and mechanical journals which were then extant. The *Proceedings of the Institution of Civil Engineers* did not appear until 1837 although a selection of papers presented before the Institution's meetings with accompanying drawings was published in three volumes in 1836, 1838 and 1842 as the *Transactions of the Institution of Civil Engineers*. Although the Institution never closed its doors to the practitioners of the specialised branches of engineering which began to emerge after the establishment of the Institution of Mechanical Engineers in 1847, the papers read before the society and published in the *Proceedings* were largely limited to subjects which are now classed as civil engineering in the narrower sense.

Mechanical engineering as a profession evolved out of the development of the rotary steam engine, the introduction of which in 1782 made the application of steam power possible to areas other than pumping. This application of steam power within industry greatly expanded in the early years of the nineteenth century necessitating the foundation of an organisation to foster the exchange of technical experience. The Institution of Mechanical Engineers was founded in 1847 'to enable mechanics and engineers in the different manufactories, railways and other establishments in the kingdom to meet and correspond, and by mutual exchange of ideas respecting improvements in the various branches of mechanical science to increase their knowledge and give an impulse to invention likely to be useful to the world'. The *General proceedings of the Institution of Mechanical Engineers* first appeared in 1847. The initial volume covered papers presented before the institution during the first three years of its existence and included items on the balancing of locomotive wheels, the prevention of explosions in boilers, the rotary engine and hydraulic starting apparatus.

The trend towards specialisation in science and engineering is further evidenced by the proliferation of societies and institutions founded in England in the remaining years of the nineteenth and the early years of the present century. Each of these bodies was to issue its own journal or proceedings. The Royal Aeronautical Society was founded in 1866, the Iron and Steel Institute of 1869, the Institution of Electrical Engineers in 1889, and the Institution of Mining and Metallurgy in 1892. Some major examples of the numerous professional institutions established in the twentieth century are the Institute of Metals, 1908, the Institution of Production Engineers, 1921 and the Institute of Welding, 1923. The pattern for the United States was similar to that in England and the other industrial European countries.

To cite some of the major American societies, the American Society of Civil Engineers was founded in 1852, to be followed by the American Chemical Society in 1876, the American Society of Mechanical Engineers, 1880, the American Institution of Electrical Engineers, 1884, the American Society for Testing and Materials, 1898, the Technical Association of the Pulp and Paper Industry, 1915 and the American Petroleum Institute, 1919.

REFERENCES

1 Phillips, J R 'Liebig and Kolbe, critical editors' *Chymia*, 2, 1966, p 89.

2 'Is it safe to look back?' *Nature*, 224, 1st Nov 1969, pp 417-22.

3 Graves, C *Life and letters of Alexander Macmillan* London, 1910.

3

Present forms of the scientific and technical journal

SEVERAL FORMS of scientific and technical periodical have evolved in the present century. These may be conveniently classified into three groups according to their publishing origin and then again within each group according to their function, thus:

a) Learned society and professional institution journals:
1 primary; 2 communications; 3 general purpose; 4 review.

b) Commercially published journals:
1 primary; 2 technical and trade; 3 controlled-circulation.

c) House journals:
1 prestige; 2 information on products; 3 internal house organs.

Learned society and professional institution journals

1 Primary. The learned society and professional institution primary journals are the main purveyors of original published science. A basic function of the scientific society is to document and disseminate information on the original research work carried out by its members. The primary journal is the chief medium of attaining this end. The societies and institutions publish papers which have usually been presented before a meeting of the body or one of its branches in an organ usually termed *Journal,* or *Transactions,* or *Proceedings.* In addition to their documentation and dissemination functions the primary journals have the role of establishing priority of scientific observation for research workers. The Council of Biology Editors in its definition of a primary publication state that 'an acceptable primary publication must be the first disclosure containing sufficient information to enable peers 1 to access observation, 2 to repeat experiments, and 3 to evaluate intellectual processes'.

It must also be 'susceptible to sensory perception, essentially permanent, available to the scientific community without restriction and available for screening by one or more of the major recognized secondary services'.[1]

The societies and institutions see themselves as the guardians of the standards of publication within their fields and it is in their journals that the majority of the original contributions to the literature of science and technology appear. Excellence of content is assured by a refereeing process undertaken by an editorial board consisting of eminent practising members of the profession or society. There are two basic reasons for this refereeing process. Firstly the journal is the official publication of the society and the board believes that the journal has an obligation to the members of the society and to the scientific community in general to ensure that the material appearing in its volumes is scientifically accurate. Secondly the process commits the author to presenting his material clearly, succinctly and consistently with certain minimal rules (house style) which will ensure standardisation of matters relating to abbreviation, bibliographical citation, etc. An editorial board will reject a paper normally on one of two grounds: a) the substance of the contribution does not meet the high standards the journal strives to maintain—the work might not be complete or the conclusions drawn might not be supported by adequate experimental evidence, b) the subject matter of the paper might also be too specialised or outside the subject scope of the journal.

2 Communications journals. The demand for a faster flow of information has fostered the 'communications' or 'letters' journal. Such organs are usually issued with a semi-monthly frequency and contain short preliminary announcements of work in progress in the manner of the letters to *Nature* and *Science*. For speed of dissemination, the communications are published in an unedited form which has caused some scientists to question the validity of part of the content of these journals. To expedite currency even further the journals are often produced by offset or other near-print processes rather than by letterpress printing. The initial function of the communications journals were to

serve as an early warning or priority claiming medium until fuller publication in a more conventional journal could be justified. But there is now evidence that scientists are using these media as a form of publication in their own right. Only fifty percent of the communications to *Physical review letters* ever appear as full scientific papers.[2]

Most of the communications journals were founded in the late nineteen fifties and early sixties. Typical examples are *Applied physics letters* (American Institute of Physics), *Chemical communications* (Chemical Society), *Electronics letters* (Institution of Electrical Engineers), *F E B S letters: for the rapid publication of short reports in biochemistry, biophysics and molecular biology* (Federation of European Biological Sciences), *J E T P letters* (American Institute of Physics) and *Physical review letters* (American Physical Society). It should be noted that the communications journal is not the sole prerogative of the society and institution, some are published commercially by scientific publishing houses as *Chemical physics letters* (North-Holland Publishing Co) and *Inorganic and nuclear chemistry letters* and *Tetrahedron letters* (Pergamon).

3 General purpose journals. Some societies and institutions will include information on their day to day activities in the primary journal but the majority will issue a 'general purpose' publication to act as the link between the body and its members. The general purpose journal may contain articles of general interest in the form of reviews of progress, items on current developments within the field and in science and technology generally, news and comment on social and economic aspects of the subject, summaries of papers to be given at future meetings, activities of society groups, book reviews, book notices, letters to the editor, information of new products, obituaries, accessions to the library and library notes. Some notable examples of these journals are *Chartered mechanical engineer* (Institution of Mechanical Engineers), *Chemical bulletin* (American Chemical Society), *Chemistry in Britain* (Chemical Society), *IEE news* (Institution of Electrical Engineers) and *Physics bulletin* (Institute of Physics and the Physical Society).

4 Review journals. Many of the primary journals contain some review papers in addition to the original papers. These summarise or evaluate the art in a highly specific sector of the subject field over a given period of time. The review will consist of a narrative of developments supplemented by an extensive list of associated bibiliographical citations to the literature. De Solla Price has given some data on the literature citation patterns of review papers. He estimates that six percent of all scientific papers are of the review type, and that these papers generate thirty seven percent of the total number of citations in the literature. Eighty three percent of review articles contain forty five or more references with an average of seventy five citations while the remaining seventeen percent contain eighty four or more references with an average of 170 citations.[3]

Reviews are of immense value to scientists who wish to familiarise themselves with the state of the art in a particular field before committing themselves to a research project. In recent years the importance of review writing has been emphasised in official pronouncements on information transfer, particularly the Weinberg[4] and SATCOM[5] reports. Both have encouraged practising scientists to undertake review writing as an integral part of their work for the benefit of the scientific community at large.

The review journal is composed entirely of review type papers or in some cases, as with *Applied mechanics reviews* (American Society of Mechanical Engineers) and *Mathematical reviews* (American Mathematical Society) of review articles and abstracts. Good examples of the purely review journal are *Chemical reviews* (American Chemical Society) and *Reviews of modern physics* (American Physical Society).

Commercially published journals

1 Primary. A substantial number of primary journals are now being marketed by commercial publishers specialising in academic and research fields. Blackwood's Scientific Publications issue amongst other titles *Journal of microscopy* and *British journal of haematology,* Taylor and Francis, *Contemporary physics* and the *Philosophical magazine,* and Pergamon Press *Talanta* and

Tetrahedron. These publications are directed towards the academic and research library—the institutional market, rather than the individual scientist who would find their extremely high subscription rates prohibitive. Thus ensured of a captive readership, the editorial policies of the journals are freed from the constrictions of normal commercial pressures. Excellence of content is achieved, as with the learned society primary journals, by the action of editorial boards composed of eminent practising members of the academic and research worlds. Membership of these boards embraces a wide representation of regional and overseas editors from the world's main research centres within the disciplines covered by the journals. The international flavour of the publications is further reflected in their multilingual content where a single journal issue will often contain papers in English, French, German and Russian. Some of the specialised publishers have cooperated with learned societies or research institutions in the publication of a primary journal. In these ventures a reduced subscription will sometimes be offered to the members of the society while non-society subscribers will be charged the full subscription. Academic Press publish *Journal of applied bacteriology* for the Society of Applied Bacteriology, Butterworth, *Combustion and flame* for the Combustion Institute, Blackwell, *Journal of food technology* for the Institute of Food Science and Technology, and Pergamon Press's extensive society programme includes *Acta metallurgica* for the American Society for Metals, the *Journal of the Franklin Institute* and *Geochimica and cosmochimica acta* for the Geochemical and Microchemical Societies.

2 Technical and trade. Most fields of industry and technology are covered by at least one commercially published technical or trade journal. These publications aim to cater for the information needs of industry by repackaging information gleaned from other, often primary sources, in an easily digestible form for management and practitioners alike. The journals' main functions are to report on new techniques within the industry and to monitor and report on new techniques developed outside the

industry which are of potential interest to the readership. The articles are usually produced by staff journalists and are invariably broader in appeal than the papers contained in the primary journals. Often one article will summarise a number of related papers from the primary literature. Some of the technical journals will however also contain quite a high proportion of specialised and highly technical articles written by commissioned authors. Good examples of long established technical journals published in the UK are *British chemical engineering*, *Electronic engineering*, *Metalworking production* and *Paint technology* and in the United States *Chemical engineering*, *Oil and gas journal*, *Power* and *Textile world*.

The major proportion of the financial income of these journals is not derived from subscriptions but from the revenue from the manufacturers whose products and services are advertised in their pages. In some fields a number of commercially published journals are in fierce competition to attract the readership with purchasing power and often the revenue accruing from advertisements will influence the editorial policy and hence the content of a commercially produced journal. It is, therefore, important for librarians and others responsible for the selection of these journals to keep them under close surveillance as a change in advertising income may substantially change their character. In addition to the longer articles, commercially published journals include regular features on new plant and equipment, processes, products and materials, editorials giving background news of the industry and letters to the editor. Other forms of literature are usually covered: trade literature will be scanned and significant items drawn to the readers' attention, new British and foreign patent specifications are sometimes abstracted and some journals provide regular coverage of abstracts of papers and articles from the world's periodical literature.

Most of the technical journals will include some commercial information relating to the industry which they serve and thus they will perform a dual technical/trade role. Trade journals are almost totally commercially oriented, their contents consisting of topical information on contracts and prices and market and company news. Their frequency is invariably weekly. Some typical examples of the genre are the British *Electrical times* and *Gas world* and the American *Chemical week* and *European chemical news*.

3 Controlled circulation. The controlled circulation journal is a particular class of commercially published periodical issued by the publisher to promote the products of the firms active in a specific industry. The numbers of these journals have increased rapidly, some would say alarmingly, in recent years as the economics of journal production have forced publishers to rely more and more on the revenue accruing from advertisements. The advertiser now pays between eighty percent and ninety five percent of the cost of most commercially published journals. Controlled circulation journals are sent regularly, at no cost, to senior technical personnel employed by those companies who are the potential purchasers of the products advertised in the pages of the journal. The reader is invited to complete prepaid postcards to obtain literature on any of the numbered items covered in the journal. The editorial content of most controlled circulation journals is usually thin. The best of the genre concentrate on interpreting technological developments rather than describing technical detail. They attempt to demonstrate how a technical development will affect the engineers' activities. Any additional information in the journals will consist of general articles summarising trends within the industry and trade fairs. A newspaper type format is often adopted by the controlled circulation journal, presumably as a ploy to persuade the reader that what he is reading *is* actually news. Most controlled circulation journals are also available on subscription to those who do not meet the requirements necessary to receive a free copy. But in these instances the subscriptions seem inordinately high as the publisher sees subscription copies as self liquidating and therefore the subscriber is expected to pay the full price for the production of the individual copy. The question of the availability of controlled circulation journals to libraries and information centres is somewhat confused. Many publishers will simply not make

their journals available to libraries on terms other than subscription. Some publishers, however, will make their journals available to special libraries if the librarians are willing to give documentary evidence that the journals are being used in the advertisers' interests. In these cases the journal should be requested by the chief engineer in the firm but addressed to the library. The librarian must then agree to route the journal directly to the engineer on its receipt. *Design & components in engineering, Fluid power international* and *Materials handling news* are examples of British publications and *Chemical processing, Electronic components news* and *Petroleum equipment news* typical American controlled circulation journals.

House journals
A house journal is a serial publication issued by an industrial, commercial, public service, or similar organisation. House journals are a form of promotional literature in that they seek to project and enhance the image of the parent organisation to its customers or employees or to advertise and promote the products and services of the organisation. The journals are usually made available on a subscription free basis to the company's potential customers or to the users of the services which the organisation purveys. It has been estimated that currently some 2,000 house journals are being published in the United Kingdom and approximately 10,000 in the United States.

There are two major categories of house journal: a) external publications directed towards the market the organisation serves and b) internal publications published as organs of communication between the company and its employees. A small number of companies publish dual-purpose journals intended for a joint company/extramural readership.

The first type of external house journal can be termed the 'prestige' publication. The purest type of this genre is the journal which promotes the image of the company not by directly publicising its wares, but more discreetly by its association with standards of excellence in both presentation and content. The outstanding example of this class of journal is ICI's exquisitely designed and illustrated *Endeavour,* published quarterly in five separate language editions as a review of progress in science and technology. The scientific content of this journal is not remotely related to its publisher's technological fields of interest—it is offered as a contribution to promoting the understanding of science. Yet another type of prestige journal is the publication which includes well-presented accounts of research or development work undertaken by company personnel, as for example the Skefco Ball Bearing Company's *Ball bearing journal* and the *IBM journal of research and development.* The most direct approach to promoting the interests of the organisation can be seen in the house journal which publishes items dealing with the utilisation of the company's products and case histories of successfully completed contracts and projects, for example *BICC news, Dow diamond* and *Dexion news.*

House journals can be a useful source of technical information but their value is not generally appreciated. In a survey of house journal usage and storage made by Isabel I Harberer[6] it was reported that 'overall there was found to be general neglect for acquiring and processing even the most informative house journals'. Their image then is very closely identified by librarians with publicity and the hard sell. This is unfortunate. House journals are invariably available freely, many of them publish articles of very real technical value, and in an industrial library they are perhaps the most convenient means of keeping abreast of the products and services of rival companies. The technical worth of the best house journals can be gauged from the fact that they are deemed worthy of coverage by the foremost abstracting and indexing services.

The 'internal' house journal is produced for a company's personnel with the intention of creating a sense of community within the organisation. It acts as a newspaper within the company, carrying news of social and sporting events, personnel changes, feature articles by employees, competitions, news of suggestion schemes and incentive bonuses, etc. The internal journal is consequently of little value as a source of technical information.

REFERENCES

1 Zwemer, R L 'Identification characteristics useful in improving input and output of a retrieval system' *Federation proceedings* 29 (5), Sept/Oct 1970, pp 1595-1604.

2 Kuney, J H 'New developments in primary journal publication' *Journal of chemical documentation* 10 (1), Feb 1970, p 440.

3 De Solla Price, D 'Networks of scientific papers' *Science* 149, 30th July 1965, pp 510-5.

4 US President's Science Advisory Committee *Science, government and information*. Washington, DC, USGPO, 1963.

5 *Scientific and technical communication: a pressing national problem and recommendations for its solution*. A report by the Committee on Scientific and Technical Communication of the National Academy of Sciences and the National Academy of Engineering. Washington, DC, 1969.

6 Harberer, I I In: *Progress in library science 1967;* edited by R L Collison. London, Butterworth: Hamden, Conn, Archon Books, 1967.

4
The scientific journal: some problems and alternatives

ALTHOUGH THE SCIENTIFIC periodical has been a major medium in the communication of scientific information for more than three hundred years it has in the past thirty years been subjected to frequent criticisms. Its purported deficiencies have been conveniently summarised in a report by Phelps and Herlin.[1] The opponents of the journal have pointed to the delays in the publication of papers of up to a year which many journals experience. The survey of the characteristics of professional scientific journals undertaken in 1962 by Campbell indicated that forty five percent of the journals examined experienced delays of more than seven months although none reported delays of more than sixteen months.[2] This delay is precipitated by deficient technology in both the production of the journal and in the management of the information packaged by editors and publishers.

A further shortcoming of the journal is the restriction often made by editors on the length of papers which causes supporting data and background information to be either omitted entirely or drastically curtailed. These constrictions are imposed because of the high costs of journal publication and the pressing claims of a large number of papers for journal space.

The dispersion or 'bibliographical scattering' of papers across a very large number of journals is a serious constraint to information gathering by scanning journals. This dispersion of papers, which is common to any given subject, renders completely comprehensive current scanning almost an impossibility to the scientist who has only a limited amount of time available for reading and whose reading patterns will anyhow normally be restricted by habit to a small number of general scientific journals and the main 'core' journals in his special subject field. An additional effect caused by scatter related to the buying of scientific journals is that the purchaser is required to pay for a large number of papers which are of little or no interest to him. Elsdon-Drew has estimated that any one article in a specialised research journal will be of interest to only ten percent of workers active in the subject field covered by the journal and that an article in a general periodical may be of interest to only two percent of the readership.[3] Similar figures have been offered by Wooster in a survey undertaken for the American Psychological Association sponsored by the National Science Foundation.[4] He estimates that half of the papers published in the core journals within the field would be read in detail by no more than one percent of their readers and that no paper would be read by more than seven percent of the readers.

Another problem associated with journal publication is inherent in the system of refereeing. Whilst most learned journals control the quality of the papers they publish by screening through an editorial board, in many cases a paper will be submitted for publication and no member of the board will be competent to pass judgment on its merit because of its depth of specialisation. In these instances the managing editor will need to invite the cooperation of an 'associate referee'. Again such suitably qualified persons can not always be located quickly and this tortuous process of refereeing will lengthen the delays experienced in publication.

Most of the criticisms launched at the scientific journal arise from the fact that the journal is a dilatory, expensive and unwieldy assemblage of papers. The alternatives that have been offered to replace the journal are in the main alternatives to the journal as a package of papers. It is a fact that the paper is vulnerable as an efficient medium of communication and dissemination of information. But it is also true that the individual paper is not the only valuable feature of the journal. The scientific periodical has responsibilities additional to that of communication. Journals form the scientific archive: they are the official public records of scientific achievement. The journal system also conveys prestige and recognition on the individual. It provides the scientist with the means to justify that his work is meaningful in the context in which he operates. Herschman considers that this function along with 'the need to believe that our intellectual effort was original (priority), has been formally validated, and is publicly known to be our own, is the social mechanism by which the community encourages us to be scientists'.[5] The serendipity factor of the usage of journals is also vital and can never be measured. The value of casual browsing through a journal could never be compensated if the form were to be completely replaced by other channels of communication. The chance discovery of a principle in one field and its application to another has been a process facilitated by the regular scanning of the scientific press.

The alternative methods of disseminating and communicating information that have been advocated to the scientific periodical are legion. The more active participation of television and radio stations has been suggested, particularly in the United States. The solution to the problems presented to librarians by the very mass of printed journals has been seen in the replacement of hard-copy journals by primary publication in microform. This method of publication has been used with some success in highly specific subject fields where the readership necessary to sustain conventional hard-copy publication is not available. The Wildlife Disease Association's journal *Wildlife diseases* has been published on microfiche only for many years and a number of microfiche only journals have been established in some specialised fields of medicine. The Institution of Electrical Engineers in 1970 became the first British learned society to use microfiche as a current publishing medium when concurrent microfiche editions of eight of its journals were instituted. The fiche editions were offered at the same published price as the printed version with an overall reduction of twenty five percent offered for a double subscription. The persisting antipathy of the scientific reader to using microforms has until recently inhibited the widespread development of this medium. This antipathy was bred on the former technological shortcomings of 35mm reading equipment. Since the late 1960s and the advent of cassette-loaded 16mm film with motorised reader-printers in which it is no longer necessary to thread the film through the machine, there has been a more ready acceptance of microfilm as a medium of journal access. The instant availability of hard-copy through the reader-printer has been an integral part of the success of 16mm film. An increasing number of learned societies are now making their journals available on 16mm film in addition to conventional publication.

Several plans to curtail the size of journals have postulated the supplementary or auxiliary publication scheme which depends on the establishment of a central depository which would hold copy of background information and tabular data too voluminous to publish economically with the original article. Such a scheme has recently been established at the British Library, Lending Division (BLLD). The library has agreed to make available on microfiche additional material such as graphs, tables, etc to supplement an article which has already been published in a scientific or technical journal. A note of the existence of the additional material is included in the parent article and the material can then be obtained by application to the BLLD.

The American Chemical Society has since 1971 utilised the microform editions of journals as a medium for supplementary primary publication. During the first year of the scheme's operation seven hundred pages of supplementary material, mostly charts and tables, from 148 papers were published solely in the microfilm editions. When the author and the journal editor have decided which additional material is to appear in the film edition the total manuscript to be published is forwarded to the editorial office. After the part of the manuscript to be published in the printed form has been page set the supplementary matter is sent for microfilm processing. Appended to the paper in the printed form is a note indicating that additional related information is available in the film edition with instructions and prices for obtaining the microfilm or a corresponding hard copy of this.

The subject rationalisation of journal publication has also been advocated. L'Hermite has suggested that journals should define their subject scope more precisely and that individual issues should be devoted to specific topics. He maintains that these arrangements could result in the individual acquiring only those papers in which he is interested.[6]

The most frequently proffered alternative to the journal has been its substitution by a system of separate papers. Under such a scheme authors would submit copies of their papers to a central editing, publishing and distributing agency. The agency would publish summary journals, abstracting the papers which it had received and these papers could then be made available on request. They could also be distributed automatically to individuals with known subject interests. Numerous such schemes have been suggested in the literature since the mid 1920s but the most widely publicised idea was that of Watson-Davies whose scheme was published as an appendix to J D Bernal's book *The social function of science*. Bernal was to become the leading protagonist of the 'separate' idea. His modification of previous plans was presented to the Royal Society Scientific Information Conference in 1948. The distinguishing feature of Bernal's scheme was that copies of the separates could be automatically distributed to libraries and documentation centres with known interests.

The call for the separate system has diminished in recent years as cogent arguments against its implementation have been presented. In fact the systems of separates that had been implemented by the American Society of Civil Engineers, the Chemical and Physical Societies of London to partially surplant journal publication were all abandoned after relatively brief periods. Separates are now seen as impractical. No scientist could possibly find time to read even a small percentage of all the papers which would be pertinent to his work. The practical problem of arranging the subject distribution of thousands of highly specific separates to a scientific community with thousands of entrance points would also indeed be daunting. In addition, what criteria could be used to estimate the number of each individual separate to be published? Another major disadvantage of the separate system would be the housekeeping problems they would present to librarians in terms of filing, binding and conservation and all the related problems of bibliographical citation and access from the point of view of the user.

A service which was designed to complement the journal and

also to expedite the dissemination of scientific ideas was pioneered by the American Mathematical Society in 1968. The scheme was conceived of as a selective dissemination of information (SDI) system. The Mathematical Offprint Service (MOS) offers to individuals on a continuing basis reprints of articles which satisfy criteria specified in user interest profiles. A contributor, in addition to denoting his subject interests through the language of the classification scheme used to arrange *Mathematical reviews*, may also include in his profile the names of authors whose work he wishes to cover or exclude and also particular journals and languages. Seventy of the journals covered by *Mathematical reviews* send galley-proofs of their papers to MOS and 450 articles are processed by the service each month. The MOS user receives his bundle of preprints each week or his computer printout of references as he prefers.

The Information Exchange Groups (IEG) experiment which flourished in the United States between 1961 and 1967 was not instituted to oust the scientific journal but rather to supplement it. The venture was mounted under the auspices of the National Institutes of Health with Federal funding under the direction of E C Albritton. The aim of an IEG was to provide a means of quick communication within the total world pool of research manpower in a sharply focused area. The groups were founded on the premise that the usual methods of exchange of information between workers in a specialised field is highly inefficient and acts as a deterrent to rapid progress. The first IEG was founded in 1961 within the field of electron transfer and oxidative phosphorylation. Membership grew to 725 scientists, 329 of these being from 32 countries outside the USA. By 1967 seven IEGs had been founded with a total membership of 3,600. The cost of supporting IEG no 1 for one year was $100,000 or approximately $140 per year per member. The groups disseminated information by members submitting communications to the group HQ in 'camera-ready' form to be duplicated by xerox and then routed to members by air-mail. The turn around time at the group HQ was said to be one week to ten days. The communications were not subjected to editorial screening as it was maintained that the judgment of one's scientific peers would act as a deterrent to the submission of marginal communications or pot boilers. Communications included: scientific papers; comments or criticisms of these papers which had been previously circulated; reports on work in progress; soss for information.

The groups had definite virtues as supplementary organs of communication to the journal. They ensured that the scientists were kept informed of all work relevant to their interests being undertaken throughout the world. They were of particular value to scientists working in laboratories in countries outside the conference circuit—the modern invisible colleges where so much information is disseminated orally. Their speed of dissemination has already been noted; it was this factor that was their main success and also the cause of their demise. The IEGs had their detractors who laid the following charges against them.

1 Their implications of selectivity. It was argued that this was improper in an operation conducted by a government agency, journals are available to all and encourage the serendipity process. But much valuable material which originated in an IEG was also subsequently published in conventional journal form.

2 It was maintained that while the IEG did accelerate communication it did not substantially add to the process as much of the information circulated was subsequently published in journal form. Albritton had given an estimate that eighty percent of the communications circulated through IEG 1 were ultimately published in the scientific press.

3 There were problems of inaccessibility: some communications were being cited in journal articles and they were then impossible to obtain for perusal. It was argued that the IEGs presented priority and piracy problems. Some maintained that unscrupulous scientists were using the communication to establish priority before theories or hypotheses had been sufficiently well formulated. Others claimed that ideas formulated in communications were being pirated and passed off without acknowledgment. Both these charges were vehemently denied by the IEGs: Albritton maintained that they were 'new and potent safeguards to priority' and that any disputes over priority or piracy

could easily be resolved by reference to the chairman of a group.

4 It was claimed that as the groups became too large the quality of the communications fell because of the absence of refereeing. The very volume of the communications would result in the IEG member being confronted with huge bundles of preprints many of which were of dubious value.

5 The high cost of operating the group was quoted—$140 per head per year. The opponents of the IEGs suggested that for this sum each member could be given free copies of the conventional journals in his field. The protagonists of the groups however insisted that their annual cost was miniscule 'related to the dividends received by the members and the total cost of supporting their research'.[7]

The experiment was wound up by the National Institutes of Health in 1967. The Institutes claimed that a) the concept had been demonstrated as workable, but b) the rapid growth of the groups had passed the point of what could be accommodated with the resources available. The experiment had been successful and Green maintained that it was 'the overwhelming success . . . which finally spelled . . . doom'. The chief reason for the demise of the groups was the antipathy of the establishment scientific press. In Vienna in September 1966 a conference of the editors of five major biological journals, the Commission of Biological Editors, representing thirteen journals in all, resolved that they had 'no wish to connive at multiple communication' and would not publish in their journals any paper which had previously been circulated through an IEG. They insisted that the activities of the groups were indistinguishable from publication in the orthodox state and made a mockery of the phrase 'personal communication'. They also refused simultaneous publication or publication through an IEG of a paper which had been published in a journal. Furthermore the editors resolved that no author of a paper appearing in their journals could be allowed to cite an IEG communication in that paper. The editors were apprehensive that the status and prestige of their journals would be diminished if the IEGs were allowed to continue distributing to their members six months to a year earlier the papers which would appear in the journals in a similar form. But the campaign against the IEGs had also been waged in the general elite scientific journals *Science* and *Nature*, which was particularly vociferous in its attack. *Nature* maintained that the defects of the IEG communications included 'inaccessibility, impermanence, illiteracy, uneven quality, lack of consideration'[8] while *Science* declared 'In an era of information explosion who needs government subsidised shoddy merchandize'.[9] In the wake of this furore the National Institute of Health terminated the experiment in 1967. Green claims that the editors of the classical journals vented on the IEGs their dissatisfaction with the present state of their journals. It was to the detriment of a section of the scientific community that the ardour of the editors' crusade against the IEGs succeeded in snuffing out these new organs of communication. The IEGs and the scientific journals could and should have complemented each other. The IEGs might have performed the roles of editorial boards of peers—a sieve through which the ideas and papers could pass and be subjected to scrutiny until the material of substance could be refined and then presented for publication for the benefit of the scientific community at large in the pages of the journal. This archive could then be made available in the libraries and information centres throughout the world.

Research newsletters have features in common with the IEG communications: they circulate privately amongst scientists, particularly in clearly defined areas within the biological sciences as a selection of their titles will indicate—*Human chromosome newsletter, Mammalian chromosome newsletter, Insect toxicologists newsletter, Mouse news*. Unlike the IEG communications they are basically *news-media* and their circulation is not normally limited. Consequently the scientific journals have never seen the newsletters as a threat to their status and prestige. The newsletters are generally available, often at a nominal cost, to those individuals and libraries who express interest. Their contents will always include social news—addresses, personal news and appointments, and in addition brief technical notes, short research reports and bibliographies. Many of the newsletters are available on loan retrospectively, the British Library, Lending Division, has a large collection and Wyatt has estimated that about ten percent of their number are listed in the *World list of scientific periodicals.*[10]

REFERENCES

1 Phelps, H R and Herlin, J P 'Alternatives to the scientific periodical' *Unesco bulletin for libraries* 14 (2), Mar/Apr 1960, pp 61-75.

2 Campbell, T H et al *Characteristics of professional scientific journals* 1962 PB 166088.

3 Eldson-Drew, R 'The library from the point of view of the research worker' *South African libraries* 23, Oct 1955, pp 51-4.

4 Wooster, H 'The future of scientific publishing' *J Washington Academy of Sciences* 60 (2), June 1970, pp 41-5.

5 Herschman, A *Journal of chemical documentation* 10 (1), Feb 1970, pp 37-42.

6 L'Hermite, R *Societe des ingenieurs civils de France* 109, Mar/June 1956, pp 229-33.

7 Green, D 'Death of an experiment' *International science and technology*, May 1967, pp 82-8.

8 *Nature*, 211 (5052), 27th Aug 1960, pp 897-8.

9 *Science*, 154 (3750), 11th Mar 1966, pp 727.

10 Wyatt, H V 'Research newsletters in the biological sciences' *Journal of documentation* 23 (45), Dec 1967, pp 321-5.

5
Access to the journal literature

1: Catalogues, bibliographies and lists of journals

NATIONAL UNION CATALOGUES

The most comprehensive listings of periodicals are the national union catalogues which record the holdings of the major national, university and institutional libraries within a country. A knowledge of their scope is essential to those involved in work with scientific and technical journals because a) their **coverage of older material** is excellent, and it is this area that the more specialised listings of scientific and technical material are usually deficient, and b) the specialist in any field is frequently obliged to refer to the literature of a related field and sometimes even to a field distant from his own. There is a difference between the periodical literature of science and the periodical literature of interest to a professional scientist. These national union catalogues are the major tools for tracing locations for sets of a particular title and they are also of value in checking bibliographical details such as changes in titles or verifying places of publication. They are of little relevance however in subject selection as they are straightforward title listings of journals with details of holdings of cooperating libraries and references between variant titles. The two main union catalogues are:

1 *British union catalogue of periodicals* 4 vols London, Butterworth, 1954, supplement 1960. The five volumes list 140,000 titles held in 440 cooperating libraries. Quarterly supplements listing new periodical titles and titles published after 1960 have been issued since 1964. These cumulate annually and quinquennially. A separate list of titles in science and technology is published each year as the *World list of scientific periodicals* (see page 60).

2 *Union list of serials in libraries of the United States and Canada* 5 vols 3rd ed New York, Wilson, 1965. Lists 150,000 titles published up to 1949 held in the Library of Congress and other

cooperating libraries. No further editions have been planned but the work has been continued as *New serial titles*, a monthly publication prepared by the Library of Congress which commenced in 1950. This has annual, quinquennial and decennial cumulations. A separate monthly edition providing a subject arrangement by the Dewey Decimal Classification is also available.

SUBJECT GUIDE

The most comprehensive and useful subject guide to periodicals published throughout the world is: *Ulrich's international periodicals directory: a classified guide to periodicals, foreign and domestic* 14th ed 2 vols New York, Bowker, 1971/72. Provides detailed information including circulation figures, coverage by abstracting and indexing services, presence of illustrations, advertisements, etc, inclusion of bibliographies, etc for some 50,000 'in-print' titles covering all fields of knowledge. Arrangement is alphabetical by title within subject groupings, *eg* astronomy, gerontology, leather, mathematics, etc. The main sequence is supplemented by a title and specific subject index. A useful companion volume to *Ulrich* is: *Irregular serials & annuals: an international directory; a classified guide to current foreign and domestic serials, excepting periodicals issued more frequently than once a year;* edited by E Koltay New York, Bowker, 1967. Arranges 14,500 yearbooks, annual reviews, advances in, progress in and similar publications under 230 alphabetical subject groupings with a supplementary title and specific subject index.

NATIONAL LISTINGS OF JOURNALS PUBLISHED WITHIN A PARTICULAR COUNTRY

National listings of titles are available for most of the major periodical publishing countries. Often a general bibliography of periodicals and a more specialised listing covering science and technology will be available for a country. These are both valuable in the subject selection of journals as they invariably incorporate subject groupings in their arrangement.

GREAT BRITAIN: GENERAL

Willings press guide London, James Willing. An annual directory giving publication details of periodicals, newspapers and annuals published in the British Isles. Arrangement is by title with a subject index under broad headings. Selective listings are given of Commonwealth and foreign titles under country only.

Newspaper press directory London, Benn Bros. An annual directory including detailed bibliographical information and notes on subject scope, editorial policy, circulation figures, etc on British newspapers, general, and trade and technical periodicals. Overseas publications are treated selectively.

Woodward, David *Guide to current British journals* 2nd ed 2 vols London, Library Association, 1973. Suitable for subject selection, arranging 4,700 current titles under subject according to the Dewey Decimal Classification. Includes information on the average number of pages per issue, circulation, coverage by abstracting services, availability of indexes and supplements. Volume two is a directory of publishers of British journals consisting of an alphabetical list of publishers together with the titles they publish and a title index.

GREAT BRITAIN: SCIENCE AND TECHNOLOGY

Martyn J and Gilchrist A *An evaluation of British scientific journals* London, Aslib, 1968. Table 1 consists of a list of 1,892 British scientific and technical journals arranged alphabetically by title. Table 2 attempts an evaluation by ranking 590 British science and technology journals according to the frequency of their citation in the 1963 and 1964 volumes of the *Science citation index* (SCI). This listing however covers chiefly the primary journals as these are the main category of publication cited by the SCI source journals. The ranking reflects only one measure of use, citation, and is again biased by SCI concentration of coverage in the pure, life and medical sciences. Table 6 consists of a subject classification of the 200 top cited journals.

UNITED STATES: GENERAL

N W Ayer & Son's directory: newspaper and periodicals Philadelphia, Ayer, 1880-. An annual directory of American and Canadian newspapers and periodicals arranged geographically under state of publication and supplemented by title and classified listings.

Faxon's librarians' guide to periodicals and American subscription catalog Boston, Faxon, 1938-. An annual gratis publication covering some 3,000 titles.

Standard periodicals directory New York, Oxbridge Publishing Co. An annual directory listing over 40,000 periodicals published throughout the United States and Canada under 200 alphabetically arranged subject headings. The main sequence is supplemented by a title index. The information given for each title includes subject scope and circulation figures. House journals are listed separately.

Special Libraries Association *Guide to special issues and indexes of periodicals;* edited by Doris B Katz and others. New York, Special Libraries Association, 1962. A useful guide arranged alphabetically by title listing those journals which regularly publish buyers' guides, directories, etc as special issues.

CANADA: SCIENCE AND TECHNOLOGY

Canada. National Science Library *Directory of Canadian scientific and technical periodicals: a guide to currently published titles* 4th ed Ottawa, 1969. A classified listing of 900 titles covering science, technology and medicine.

CHINA: SCIENCE AND TECHNOLOGY

Great Britain. National Lending Library for Science and Technology *List of scientific and technical periodicals received from China* Boston Spa, 1964. A listing of 155 titles held by the library arranged under English/Latin/transliterated title followed by Chinese title.

Shih, B P N and Snyder, R L *International union list of communist Chinese serials, scientific, technical and medical, with selected social science titles* Cambridge, Mass, Massachusetts Institute of Technology, 1963. Part A is a listing of 500 periodicals with holdings of 28 American, Canadian, British and Japanese libraries.

Wu, J *Chinese scientific and technical serial publications in the collections of the Library of Congress* 1961 Washington, Library of Congress Science & Technology Division. Lists 1,100 titles under seven broad subject groupings with further subdivisions.

FRANCE: GENERAL

Annuaire de la presse française et etrangere et du monde Paris, Annuaire de la presse. An annual classified arrangement of French periodicals with highly selective listings of titles from other countries.

FRANCE: SCIENCE AND TECHNOLOGY

Union Internationale des Editeurs-Exportateurs de Publications Françaises *Catalogue des publications françaises: scientifiques, techniques, professionnelles, agricoles,* 1964-65 Paris, 1964. A listing of 600 periodicals with bibliographical information and descriptive annotations in English, French and Spanish.

GERMANY: GENERAL

Deutsche presse 1961: zeitungen und zeitschriften Berlin, Duncker & Humbolt. A broad subject grouping of 6,500 West and East German titles supplemented by a title index.

Stamm, W *Leitfaden für presse und werbung* Essen-Stadtwald, Stamm-Verlag GmbH, 1967-. An annual directory listing East and West German newspapers and periodicals under title and place of publication.

GERMANY: SCIENCE AND TECHNOLOGY

Saarbuch, W E *Subscription catalogue* Berlin, Saarbach GmbH. An annual listing arranged alphabetically by title, with a subject index, of German scientific, technical and medical periodicals.

Sticker, B *Verzeichnis deutscher wissenschaftlicher zeitschriften,*

1968 Weisbaden, 1968. Lists 2,000 German scientific titles within 25 subject groups.

JAPAN: GENERAL

Catalogue of Japanese periodicals. Tokyo, Maruzen. An annual listing of periodicals within subject groupings. Titles are given in transliterated Japanese or English and a title index is provided.

JAPAN: SCIENCE AND TECHNOLOGY

Japan. National Diet Library and Ministry of Education *Directory of Japanese scientific periodicals* 1967 Tokyo. Lists 5,000 current scientific and technical titles by UDC with a Japanese/English title index.
United States. Library of Congress. Reference Department. Science and Technology Division *Japanese scientific and technical serial publications in the collections of the Library of Congress* Washington, 1962. Lists 1,700 titles in two sequences: a) Western language titles, b) Japanese titles.

USSR: GENERAL

Gazety i zhurnaly SSSR Collets Holdings. An annual classified listing of Russian current periodicals available for purchase through Collets Holdings. Titles are given in Russian, in transliteration and in English, with title indexes in Russian, English, French, Spanish and German.

USSR: SCIENCE AND TECHNOLOGY

Mezhenko, Yu A *Russkaya tekhnicheskaya periodica 1800-1916. Bibliograficheskii ukazatel'* Moscow, Akademii Nauk SSSR, 1955. An alphabetical title listing, with classified, geographical and chronological indexes, of 400 periodicals published between 1800 and 1916.
Zikeev, N T *Scientific and technical serial publications of the Soviet Union, 1945-1960* Washington, Library of Congress, 1963. A title listing of 5,000 items held by the Library of Congress.

BIBLIOGRAPHIES OF SCIENTIFIC AND TECHNICAL PERIODICALS

The bibliographies of scientific and technical periodicals can be grouped into general and then specific subject listings. The most comprehensive guide to the bibliographies and catalogues of scientific periodicals is:

Fowler, M J *Guides to scientific periodicals: annotated bibliography* London, Library Association, 1966. This work covers 1,060 publications and is arranged by UDC within three main sections: 1 Universal guides (*ie* those covering journals of all countries): a) general, b) special subject lists; 2 Guides to the periodicals published by international organisations; 3 national and other regional guides. An excellent author/title/subject index is provided.

BIBLIOGRAPHIES OF EARLY SCIENTIFIC PERIODICALS

There are two important general catalogues of early scientific periodicals:

Bolton, H C *Catalogue of scientific and technical periodicals, 1665-1895, together with chronological tables and a library checklist* 2nd ed Washington, Smithsonian Institution, 1897 (Smithsonian Miscellaneous Collections vol 40, no 1076). An alphabetical title catalogue under earliest title arranged in two sequences the first being a corrected reprint of the 1885 edition containing 4,954 titles and the second a supplementary listing of 3,600 new titles. The preface states that 'it is intended to contain the principal independent periodicals of every branch of pure and applied science, published in all countries from the rise of this literature to the present time'. A complementary volume to Bolton is:

Scudder, S H *Catalogue of scientific serials of all the natural, physical and mathematical sciences, 1633-1876* Cambridge, Mass, Harvard University, 1879. Reprinted New York, Kraus Reprints, 1965. Arranges 4,400 titles by country then by place of publication with indexes of town, title and subject.

Fielding H Garrison's 'Medical and scientific periodicals of the seventeenth and eighteenth centuries' in the *Bulletin of the*

Institute of the Johns Hopkins University Vol 2 Part 5, July 1934, pp 285-343 records some 1,200 titles published up to 1800.

The most comprehensive lists of contemporary scientific and technical periodicals are again the catalogues of the largest national science collections. As with the more general listings their chief value is in bibliographical checking and the location of a title.

Great Britain. National Lending Library for Science and Technology *Current serials received by the NLL,* March 1971. London, HMSO 1971. Arranged in three alphabetical title sequences: 1 Current serial titles except for Cyrillic; 2 Cyrillic titles; 3 Cover to cover translations of Cyrillic serials. Lists 36,000 titles including social science serials and some titles from the humanities.

Great Britain. National Reference Library of Science and Invention. *Periodical publications in the National Reference Library of Science and Invention* 3 vols London, British Museum, 1969-1970. Pt 1: *List of non-Slavonic titles in the Bayswater Division* 1969. Lists 8,000 titles in the life sciences. Pt 2: *List of Slavonic and East European titles in the Bayswater Division* 1970. Lists 2,500 titles. Pt 3: *List of current titles in the Holborn Division* 4th ed, 1970. 9,000 titles relating to the technologies and the sciences on which they are based. Three alphabetical title listings supplemented by *Science Reference Library periodical news,* 1970-. A monthly list of changes of title, deaths, amalgamations and additions to stock.

Great Britain. Science Museum. Library *Periodicals on open access* 1973. An alphabetical title listing.

United States. Library of Congress. Science and Technology Division *A list of scientific and technical serials currently received by the Library of Congress* Washington, Library of Congress, 1960. A catalogue of 13,000 journals arranged in two sequences: a) those published in countries using Latin, Cyrillic and Greek alphabets, arranged alphabetically by title, and b) Oriental journals arranged under country.

In addition to these individual library catalogues the following tools should be noted:

International Federation for Documentation *Technical journals for industry* The Hague, 1967-. Separate country listings are available for: United Kingdom, The Hague, 1970; Germany, Frankfurt am Main, 1970; Czechoslovakia, Prague, 1969; Poland, Warsaw, 1967; Hungary, Budapest, 1972; France, Paris, 1969; Italy, Rome, 1967; Finland, Hensinki, 1967; Norway, Oslo, 1968; Sweden, Stockholm, 1969; Denmark, Copenhagen, 1967; Netherlands, The Hague, 1967; Turkey, Ankara, 1969; South Africa, Pretoria, 1967; Canada, Ottawa, 1967; Brazil, Rio de Janeiro, 1969; Indonesia, Djakarta, 1968; Australia, Melbourne, 1968. Each list provides information on subject coverage, readership, content and character, circulation, number of pages per issue and price.

World list of scientific periodicals published in the years 1900-1960 3 vols 4th ed London, Butterworth: Hamden, Conn, Archon Books, 1963/66. Lists 60,000 titles with the holdings of 300 libraries. The main work is supplemented by *New periodical titles, 1960-68 volume* London, Butterworth, 1970. No further editions will appear in the same format but the work is being continued in the annual supplements of the *British union catalogue of periodicals* as *World list of scientific periodicals, scientific, medical and technical entries from the British union catalogue of periodicals*.

Brown, C H *Scientific serials; characteristics and lists of most cited publications in mathematics, physics, chemistry, geology, physiology, botany, zoology and entomology.* Chicago, Association of College and Reference Libraries, 1956. (ACRL Monograph no 16.) A citation analysis of scientific serials with lists of the most frequently cited journals in eight fields arranged in descending order of citation.

National Federation of Science Abstracting and Indexing Services *A list of serials covered by members of the NFSAIS* 2 vols 1962. An alphabetical listing of 17,000 titles covered by member services, *Chemical abstracts, Engineering index, etc* with an indication for each title of the covering services.

SPECIFIC SUBJECT LISTINGS OF PERIODICALS

Guides to the periodicals covering a specific field consist of a) catalogues of the holdings of the main special libraries or national collections within the field; b) union lists of holdings of a number of libraries; and c) lists of periodicals covered by abstracting and indexing services. These lists are normally published in each issue of the abstracting journal but when the coverage is extensive, *eg* that of *Chemical abstracts,* the list is published separately with detailed bibliographical information, thus becoming an important bibliographical tool. A selection of the three categories of guides is given below under specific subject headings with brief annotations indicative of arrangement and the presence of indexes and where possible the number of titles listed. In addition to these listings, the numerous subject guides to the literature of specific subjects which have appeared in recent years, *eg* Burman, C R *How to find out about chemistry,* Tapia, E *Guide to metallurgical literature,* are often useful in subject selection as they will usually include a section on periodicals. A checklist of these guides has been compiled by the Queens University of Belfast, Science Library, Information Service *Information bulletin: guides to subject literature.*

AERONAUTICAL ENGINEERING

Royal Aircraft Establishment *Catalogue of periodicals; compiled by D I Raitt* Farnborough, 1969. 1,900 entries; alphabetical by title; subject index.

United States. Library of Congress. Science and Technology Division. *Aeronautical and space serial publications* Washington, United States Government Printing Office, 1962. 4,500 titles; arrangement under country, then by title; title index.

AGRICULTURE

Boalch, D H *Current agricultural serials: a world list of serials in agriculture and related subjects (excluding forestry & fisheries) current in 1964* 2 vols Oxford, International Association of Agricultural Librarians and Documentalists, 1965/67. Vol 1: 12,500 titles arranged under broad subject; vol 2: subject and country indexes.

International Institute of the Cooperation of Agricultural Research *List of publications, reviews and bulletins in the agricultural research field.* 700 titles arranged alphabetically by country.

Organisation for European Economic Cooperation. European Productivity Agency. *List of agricultural press and periodicals in OEEC member countries.* Paris, OEEC, 1960. 1,600 titles arranged alphabetically by country then by subject.

United States. Department of Agriculture *List of serials currently received in the library of the Department of Agriculture as of July 1 1957* Washington, United States Government Printing Office, 1958. 20,000 entries arranged alphabetically by title.

ANTHROPOLOGY

Royal Anthropological Institute. *Survey of anthropological journals and monograph series in libraries in the United Kingdom* 1957. 650 titles in thirty five university and society libraries are listed alphabetically by title.

BOTANY

Lawrence, G H M et al. *B-P-H: botanico-periodicum-Huntianum* Pittsburgh, Hunt Botanical Library, 1968. 25,000 titles relating to the plant sciences and botany are listed.

London University. Library *Botanical periodicals in London* 1954. 471 titles held in thirty two libraries are listed alphabetically by title.

Royal Botanic Gardens *Periodicals listed by first title page with references from alternative titles* Kew, 1971. 3,000 titles.

CHEMISTRY

Chemical abstracts service source index New York, American Chemical Society. An alphabetical list of over 21,000 titles providing bibliographical information on journals covered by *Chemical abstracts* or cited in *Beilstein's handbuch der organischen chemie and Chemisches zentralblatt* giving holdings of 325 American and

72 overseas libraries. The latest edition consists of all data accumulated to September 1969 and this is supplemented by the *CAS source index quarterly* which reports new bibliographical and library holdings data.

Pflücke, E H Maximilian and Hawelke, A *Periodia chimica: verzeichnis der im chemischen zentralblatt referierten zeitschriften mit den entsprechenden genormten titelabkürzungen* 2nd ed Berlin, Akademie-Verlag, 1961. Supplement 1962. An alphabetical title listing extending into the fields of technology and medicine. 8,000 journals are covered in the two volumes.

COMPUTERS

National Computing Centre *A world list of computer periodicals*. Manchester, 1970.

ELECTRICAL ENGINEERING

Central Electricity Generating Board *Union list of periodicals held by the libraries of CEGB* London, 1970. 2,000 titles listed alphabetically by title.

ELECTRONICS

ASLIB Electronics Group *Union list of periodicals on electronics and related subjects* London, 1961. 1,500 titles held in sixty libraries are listed alphabetically by title.

ENGINEERING

Engineering Index, Inc *Publications indexed for engineering (PIE)* New York. An annual title listing of journals covered by *Engineering index*.

ENTOMOLOGY

British Museum (Natural History). Department of Entomology *A list of serial publications in the library of the Department of Entomology* 2nd ed London, 1962. 1,000 journals are listed alphabetically by title.

FOOD TECHNOLOGY

Baker, E A and Foskett, D J *Bibliography of food: a select international bibliography of nutrition, food and beverage technology and distribution, 1936-1956* London, Butterworth, 1958. Arrangement under broad subject grouping with each further subdivided. Within each group there is an alphabetical list of periodicals.

GEOLOGY

Geological Society of London. *List of serials currently taken by the library* London, 1962. 550 titles arranged alphabetically by title.

Geoserials 1969. Geoscience documentation vol 1 no 1, July 1969. London & Calgary, Lea Associates, 1969. An alphabetical list of 1,500 titles with brief bibliographical details.

MATHEMATICS

London University. Library *Union list of periodicals on mathematics and allied subjects in London libraries* 2nd ed. 1968. An alphabetical list of 920 titles held in forty five libraries.

Steeves, H A *Russian journals of mathematics: a survey and checklist* New York, Public Library, 1961. Lists 250 titles in three sequences according to the number of papers published in an average issue: 20+, 7/20, 3/6.

A select list of British scientific periodicals: mathematics London, Royal Society, 1962. An alphabetical list of thirty three titles.

MECHANICAL ENGINEERING

Houghton, Bernard *Mechanical engineering: the sources of information* London, Bingley: Hamden, Conn, Linnet, 1970. Chapter 10 is a subject directory of mechanical engineering periodicals which arranges 750 titles under alphabetical subject headings. The main sequence is supplemented by a title and specific subject index.

MEDICINE

British Council and Royal Society of Medicine *Select list of British medical periodicals: an annotated guide* London, 1963. An alphabetical list of 162 titles supplemented by a subject index.

LeFanu, W R *British periodicals of medicine: a chronological list* Baltimore, Johns Hopkins Press, London, Oxford University Press, 1938. A chronological survey of medical periodicals issued in all British lands from the seventeenth century to the end of the nineteenth.

London School of Hygiene and Tropical Medicine. *Serials catalogue* Boston, Mass, G K Hall, 1965. A photolithographed listing of 6,000 catalogue cards.

Medical Library Association *Periodicals and Serial Publication Committee. Vital notes on medical periodicals: five-year cumulated index, 1952-1957* Columbia, Missouri, 1958. Brief bibliographical details of 3,500 journals arranged alphabetically by title.

Medical Libraries Center of New York *Union catalogue of medical periodicals* 3rd ed New York, 1969.

United States. National Library of Medicine *Medical serials 1950-1960: a selective list of serials in the National Library of Medicine* Washington, US Government Printing Office, 1962. An alphabetical list of 9,000 titles.

World medical periodicals 3rd ed New York, World Medical Association, 1961. An alphabetical list of 5,000 with brief bibliographical details. Subject and country indexes are appended.

MILITARY SCIENCE

United States. Department of the Air Force. Air University Library *Union list of military periodicals* Maxwell, Alabama, 1960. An alphabetical list of 1,000 titles held in thirty nine libraries.

MINERALOGY

British Museum (Natural History). *List of serial publications in the library of the Department of Mineralogy* London, 1959. An alphabetical list of 7,000 titles.

NATURAL HISTORY

British Museum (Natural History) Library *List of serial publications in the British Museum (Natural History) Library* London,

1969. An alphabetical listing of 12,500 titles held by the libraries of the Department of Botany, Entomology, Mineralogy, Palaeontology and Zoology.

NUCLEAR SCIENCE

International Atomic Energy Agency *List of periodicals in the field of nuclear energy* Vienna, 1963. An alphabetical list of 520 titles supplemented by a country index.

United States Atomic Energy Commission. Division of Technical Information *Serial titles cited in 'Nuclear science abstracts'* Oak Ridge, Tenn, 1970. An alphabetical listing of 7,000 titles.

PHYSICS

American Institute of Physics *The periodical literature of physics: a handbook for graduate students, by Robert E Maizell and Frieda Siegel* New York, 1961. A survey of the literature of physics with sections covering research, review and abstracting and indexing journals.

RAILWAY ENGINEERING

British Transport Commission *British Railways Research Department Library*. Stocks of periodicals held in the library at London Road, Derby and 222 Marylebone Road, London, NW1 Derby, 1960. An alphabetical listing of fifty five titles.

REFRIGERATION ENGINEERING

International Institute of Refrigeration *Bibliographical guide to refrigeration, 1953-1960* Oxford, Pergamon Press, 1962. Includes a list of 1,100 periodicals arranged alphabetically under country then under title.

RUBBER AND PLASTICS

University of Akron. Rubber Division Library *Union list no 5 of serials relating to the fields of rubber, resins, plastics and textiles held by the technical libraries of the Columbia-Southern*

Chemical Corporation Akron, Ohio, 1962. An alphabetical listing of 500 titles.

TEXTILES

ASLIB. Textile Group *Union list of holdings of textile periodicals* 3rd ed London, 1962. An alphabetical listing of 500 titles.

ZOOLOGY

British Museum (Natural History) *List of serial publications in the libraries of the Departments of Zoology and Entomology* London, 1967. An alphabetical listing of 3,500 titles.

Royal Society *A select list of British scientific periodicals: zoology* London, 1963. An alphabetical list of 50 titles with brief bibliographical details.

HOUSE JOURNALS

British house journals are listed alphabetically under publishing firm in the *British Association of Industrial Editors yearbook,* London, which is equipped with a title index but which unfortunately lacks both subject and industry indexes. An additional useful source of information giving descriptive notes on British house journals is the directory by Isabel J Harberer contained in *Progress in library science, 1967;* edited by R L Collison, London, Butterworth: Hamden, Conn, Archon Books, 1967, pp 17-96.

US and Canadian house journals are listed every three years in the *Gebbie house magazine directory,* House Magazine Publishing Co Inc, Sioux City, Iowa. This alphabetical listing of firms publishing house organs is supplemented with town, title and industry classification indexes.

6
Access to the journal literature

2: Abstracts—their nature and use

AN ABSTRACT can be defined as a summary of the information in a document accompanied by an adequate bibliographical citation to enable the document to be traced. The abstract is probably the most convenient method of summarising and recording for both current and future use the information contained in the articles and papers in the journal literature. Two essential features of an abstract are firstly, it should, while being grammatically correct, be as brief as possible—it will, however, be seen later that other factors will determine the type and therefore the ultimate length of the abstracts in a given journal. Secondly, the abstract should generally be written in an objective and uncritical manner—although it must be remembered that the author of the article being abstracted may himself be critical of a theory or a hypothesis. The opinions of the abstractor should not impede the flow of information between the author of the paper abstracted and the reader of the abstract. Collison identifies the task of the abstractor nicely, it is 'to convey what the author himself has done, why he has done it, and by what steps he has arrived at his conclusions, together with these conclusions. Any other points are irrelevant.'[1]

The generalised definition of an abstract given above requires some amplification as there are different degrees of abstracting but basically all abstracts are either *indicative* or *informative*. The *indicative abstract,* or as it is alternatively known the *descriptive abstract,* is a brief abstract to indicate the scope and content of a document. The *informative abstract* is an abstract summarising the principal arguments giving the principal data contained in the original document. The indicative abstract then disseminates *awareness* of information while the informative abstract actually disseminates the essence of the information and to some measure serves as a substitute for the original paper by condensing the major conclusions and presenting the significant methods and data. An indicative abstract relating to a paper on information retrieval may contain the observation 'recall and precision figures were given for the MEDLARS system' while the informative abstract would state 'the MEDLARS system was found to be operating at eighty six percent recall and sixty percent precision'. Excellent examples of services providing good informative abstracts are *Chemical abstracts* and *Biological abstracts*. *Engineering index* and the various specialised sections of the French *Bulletin signaletique* serve well as examples of the indicative type. But not all abstracting services have a rigid policy of providing either indicative or informative abstracts, some will vary the depth of information given in each abstract according to the availability of the original document and the accessibility of its contents. Articles in journals from Eastern European or Japanese sources would be given informative treatment as the originals may a) not be easily obtainable and b) even when they can be obtained their contents may be unintelligible to the vast majority of readers because of the language problem.

The *slanted abstract* is informative and may be defined as an abstract giving emphasis to a particular aspect of a document to cater for the specialised interests of a particular group of readers. It may be that the topic highlighted in the slanted abstract is not the major feature of the article, *eg* a paper describing a new analytical technique may concentrate on the methodology used but it may also describe the glassware used in the instrumentation—if this paper was being abstracted for *Glass abstracts* the concentration would be upon the nature of the glassware—the analytical technique, the major point of the article, would be of little interest to the users of *Glass abstracts*. The abstract of the same paper appearing in *Analytical abstracts*, a journal for chemists, would be substantially different in emphasis.

In the opening paragraph it was stated that abstracts should be uncritical; this is a general rule which applies to abstracts intended for all the scientific community or particular sections

REFRACTORY MATERIALS—Contd.

| | ≈(100), and [001]| | ≈[001], were observed. The observed habit planes can be predicted using a single Bain correspondence. 22 refs.

Bansal, G.K. Case Western Reserve Univ, Cleveland, Ohio; Heuer, A.H. *Acta Metall* v 22 n 4 Apr 1974 p 409-417.

Alumina See COKE OVENS—Refractory Materials; MOLYBDENUM AND ALLOYS—Dispersion Hardening.

Bonding

043938 CHEMICALLY BONDED REFRACTORIES—A NEW ERA. The manufacture of an unburned or chemically bonded refractory brick requires essentially the same processing as a burned brick, but with two important exceptions: addition of a permanent chemical bonding agent during the mixing and tempering stages; and the elimination of the burning stage. Paper presents a comparison between a burned magnesite-chromite brick and a sodium polyphosphate magnesite-chrome brick with both types of brick having identical magnesite and chromite compositions. The comparison shows that a new generation of chemically bonded brick is here and they have the advantages of having in many instances, higher cold and hot strength, lower thermal conductivity, greater volume stability, more precise shape, and due to the elimination of the burning during manufacture—are more readily available than their burned counterparts.

Geisler, Thomas A. Lavino Div, IMC. *Rock Prod* v 77 n 5 May 1974 p 85-87, 126-127.

Hydrolysis See ALUMINA.

Magnesia

043939 DEFECTS IN OXIDES. The purpose of this paper is to discuss the physical and chemical properties of the oxides are influenced by the presence of these defects. Methods of expressing the types and concentrations of defects is discussed. Notation for defects, the application of the mass action law and Brouwer and Kroger-Vink diagrams is reported. The relation between the oxygen or metal partial pressures in equilibrium with the oxide and the number and type of defects present is discussed. 34 refs.

Wagner, J. Bruce Jr. Northwest Univ, Evanston, Ill. *Am Ceram Soc Bull* v 53 n 3 Mar 1974 p 224-231.

Magnesite See GLASS MANUFACTURE—Melting.

Manufacture

043940 REFRACTORIES INDUSTRY: ITS RELATIONSHIP TO THE U.S. ECONOMY AND ITS ENERGY NEEDS. Using a 127 by 127 sector input-output matrix. It was determined mathematically that: 1) the refractories industry has a direct or indirect impact on every other sector of the industrial complex, 2) each dollar of refractories output supports $1266 of GNP and 3) the least damage of GNP occurs when energy supplies are allocated to industry in an even-handed, nondiscriminatory fashion.

Ayers, Ronald F. Battelle, Columbus Lab, Ohio; Barr, Harry W. Jr.; Fisher, W. Halder; Duckworth, Winston H.; McCoy, Larry G. *Am Ceram Soc Bull* v 53 n 3 Mar 1974 p 220-223.

Thermal Conductivity

043941 K VOPROSU OB OPREDELENII NELINEINOI ZAVISIMOSTI KOEFFITSIENTA TEPLOPROVODNOSTI OT TEMPERATURY. [Problem of Determination of the Nonlinear Dependence of Thermal Conductivity Coefficients on Temperature]. A method is proposed for the determination of the dependence of the thermal conductivity of refractory dielectrics on temperature $\lambda(t)$ at high temperatures. It is assumed that measurements of the heat flux are implemented with a random relative error having constant dispersion. The method can easily be modified for the statistical hypotheses of the like type. The method presented is realized in the form of a program in ALGOL-60 language. An evaluation of the systematic error of determination of $\lambda(t)$ by means of the conventional method is carried out with respect to the characteristic materials based on Al_2O_3 and MgO. While the conventional method shows errors at high temperatures, the method proposed in this paper shows excellent agreement with experimental data. 10 refs. In Russian.

Zborovskii, I.D. All-Union Refract Inst, USSR; Zeliger, G.Ya.; Litovskii, E.Ya. *Teploenergetika* n 3 Mar 1974 p 70-71.

Zirconia

043942 MICROSTRUCTURAL DEVELOPMENT IN PARTIALLY STABILIZED ZrO_2 IN THE SYSTEM CaO-ZrO_2. The microstructures of 3 zirconias partially stabilized with CaO were investigated using scanning electron microscopy and qualitative and quantitative X-ray analysis. The structure was closely related to the heat treatments involved in fabrication. A bimodal structure with small grains of pure ZrO_2 dispersed along the grain boundaries of larger cubic solid-solution grains developed during slow cooling from 1850° to 1300°C. The presence of a liquid phase greatly enhances the growth of the pure ZrO_2 phase. An anneal at 1300°C induces precipitation of fine ZrO_2 particles within the solid-solution grains. The relative mechanical strengths of the materials are explained in terms of the weakening of the grain boundaries associated with the transformation of the grain-boundary phase on cooling. 14 refs.

Green, David J. McMaster Univ, Hamilton, Ont; Maki, Dennis R.; Nicholson, Patrick S. *J Am Ceram Soc* v 57 n 3 Mar 1974 p 136-139.

REFRACTORY METALS See Also ELECTRIC ARCS.

Brazing See AIRCRAFT MATERIALS—Brazing.

Mechanical Properties

043943 O SVYAZI KRATKOVREMENNYKH MEKHANICHESKIKH KHARAKTERISTIK S DLITEL'NOI PROCHNOST'YU. [Relation between Short-Time Mechanical Characteristics and Long-Time Strength]. Some general relations of dependence of the short-time mechanical characteristics on the long-time ones for most of the heat-resistant materials are considered. It is shown that in the range of stresses $10 < \sigma_t < 60$ kgf/mm² there exists a connection, interesting from the scientific and practical points of view, between the short-time mechanical characteristics and the limit of long-time strength with the service life $t = 10,000$ h. 9 refs. In Russian.

Krivenyuk, V.V. Acad of Sci of the Ukr SSR, Kiev. *Probl Prochn* v 6 n 3 Mar 1974 p 31-35.

Protective Coatings See NICKEL AND ALLOYS—Protective Coatings.

REFRIGERATING MACHINERY

Compressors See Also SEAWATER—Salt Removal.

043944 EINSATZGRENZEN OFFENER EINSTUFIGER KOLBENKOMPRESSOREN. [Utilization Limits of Open, Single-Stage Piston Compressors]. Information is given on the careful measurement of pressure valve temperature at individual cylinders of capacity controlled compressors and on the obtained permissible deepest evaporation temperatures for full and part load operation with R22. At the same time, thorough investigation into the influence of the cut-off sequences of the cylinders on the obtainable lowest evaporation temperature and the degree of cyclic variations is undertaken. Distinct specific weights of the different refrigerants and utilization of ranges, and the variable gas velocities at different numbers of revolutions need adjustment of the stroke and springs of the operating valves. The criteria for defining operating ranges of the applied valves are discussed. In German.

Spott, Karl Heinz Sulzer Escher Wyss, Lindau, Ger. *Klim Kaelte Ing* v 2 n 4 Apr 1974 p 143-146.

REFRIGERATION See HEAT TRANSFER—Vapors; VAPORS—Condensation; THERMOSTATS—Refrigeration; HEAT PUMP SYSTEMS.

REFUSE DISPOSAL

043945 TOTAL CONCEPT SYSTEM FOR MUNICIPAL WASTE DISPOSAL. A review of the past and present means of waste disposal practices is given with individual evaluations of each. The negative vs the positive mode of thinking with respect to waste disposal uses are discussed. Based on the positive aspect element, two total concept systems are technically developed. Existing and proposed useful end product concepts are examined and evaluated. Continuity is maintained by proceeding to examine the economic aspect of the respective systems proposed and how each relate, in its resultant cost estimate and to the ultimate financial impact on the community, to a system's capacity as well as on a per capita basis. A summary of the success of the proposed systems to the basic criteria, as outlined in the "positive aspect" approach to the disposal problem, is given for each of the eight conditions initially cited. 26 refs.

Nagel, Lester L. Environ Prot Agency-Reg II, New York. NY. *ASME Natl Incinerator Conf, Proc, Pap, Miami, Fla, May 12-15 1974* p 33-41. Publ by ASME, New York, 1974.

043946 INVESTOR-OWNED PUBLIC UTILITIES FOR MUNICIPAL REFUSE DISPOSAL. The public's insistence on avoiding air and water pollution and achieving resource recovery challenge the tradition of construction and operation of disposal facilities wholly by municipalities. A modern energy-producing refuse incinerator meeting federal air standards requires an operating team with technical abilities and salary scales beyond those of current civil service standards. Similarly, the skills needed for marketing recovered materials find no parallel in usual municipal functions. If the refuse is hauled to remote areas, control by the originating city becomes even more difficult. When these emerging situations are coupled with the burgeoning of aggressive stock-holder-owned companies in municipal refuse work, a strong base is present for the public utility approach to the design and management of the service. Recent New York State Legalization of long term disposal contracts recognizes this. Prospects and problems which lie ahead are the subject of this paper.

Wegman, Leonard S. Leonard S. Wegman Co, New York. NY. *ASME Natl Incinerator Conf. Proc. Pap, Miami, Fla, May 12-15 1974* p 227-235. Publ by ASME, New York, 1974.

Composting

043947 POWER REQUIREMENTS OF A COMPOST CHANNEL FOR ANIMAL WASTES. A compost channel was constructed using 1-in. by 4-ft by 8-ft exterior plywood sheets for the sides and bottom. The top edge of the sides was rabbeted to receive a flush-mounted 2-in. angle iron. Supports were prefabricated using TECO 2½-in. split-ring connectors and were placed 2 ft on center. The reseach channel was elevated 3 ft to permit modifications to the channel bottom. Experimental design and procedure are described. 6 refs.

Hummel, J.W. Univ of Md, College Park; Schwiesow, W.F.; Willson, G.B. *Trans Am Soc Agric Eng (Gen Ed)* v 17 n 1 Jan-Feb 1974 p 70-73.

043948 POTENTIAL OF SPENT TEA LEAVES FOR ANIMAL FEEDS AND COMPOSTING. The chemical composition of untreated spent tea leaves is presented. The nitrogen (N) content, and the calculated crude protein equivalent (26.4%) suggests that tea waste may be usable as a protein supplement. However, the

411

FIGURE 8: *Examples of indicative abstracts from* Engineering index. (Copyright © Engineering Index Inc 1974.)

WILDNER and G. WOLFF. (Zentralinst. Krebsforsch., Akad. Wiss., Robert-Roessle-Klin., Lindenberger Weg 80, 1115 Berlin-Buch, E. Ger.) Vorschlag zur pathologisch-anatomischen Klassifikation des Brustdruesenkrebses nach dem TNM-System. [Proposal for pathological classification of mammary carcinoma according to the TNM system.] ARCH GESCHWULSTFORSCH 41(2): 146-163, 1973. [In Ger. with Engl. summ.]--A prospective study to evaluate the clinical classification, proposed in 1959, of mammary carcinoma according to the tumor node metastasis (TNM) system was done with 619 patients from 1960-1966. Analyzing the relations existing between clinical TN categories and the pathological-anatomical findings (size and extent of the primary tumor, histologically proved metastases in the axillary lymph nodes) and examination of 5 yr survival rates in relation to the clinical and pathological-anatomical finding led to the following conclusion. With rising TN category (from T1—T4, from N0—N3) the occurrence of locally advanced and big tumors and the extent of regional lymph node metastases increases. Deviations are rather large in some cases. Palpation is not reliable. In category N0 and category N1 a (i.e. palpable movable axillary lymph nodes regarded as free of cancer) about 55% were free of cancer even when examined histologically. Category T1 (palpable tumor diameter smaller than 2 cm, no retraction of the skin and/or mamilla) and T2 (palpable tumor diameter below 5 cm or below 2 cm with signs of skin retraction) give the same survival rates. Retraction signs are important for diagnosis, but have no influence on prognosis and are dispensible for clinical classification of the tumor category. This was taken into account in the revised version of the TNM classification of May, 1971. The application of the proposed staging scheme in 552 patients resulted in the distribution and survival rates li...

androgenic endocrine function is described. The tumor was dominated by heterotopic tissue composed of mucous-secreting acini with argentaffin cells present in the lining epithelium. Studies of the tumor by electron microscopy showed the columnar cells of the acini to have the structure of intestinal epithelium and confirmed the histological demonstration of argentaffin cells. This offers support for the theory that occasional androblastomas may develop as variants of a teratoma.
--E. S.

21205. ISENBERG, JON I.*, JOHN H. WALSH and MORTON I. GROSSMAN. (Gastroenterol. Sect., Veterans Adm. Wadsworth Hosp. Cent., Los Angeles, Calif. 90073, USA.) Zollinger-Ellison syndrome. GASTROENTEROLOGY 65(1): 140-165, 1973.--This selective review emphasizes recent developments.. It is now unequivocally established that Zollinger-Ellison syndrome (ZES) is caused by a tumor of the pancreas or duodenum that secretes excessive amounts of gastrin. Many patients with ZES probably remain undiagnosed because their symptoms do not distinguish them from patients with ordinary peptic ulcer disease. Most reported cases probably represent the more flagrant forms of the disease. It would be incorrect to assume that the cases reported in the literature are a random sample of the entire population of patients with ZES. ZES accounts for less than 1% of peptic ulcer disease. Until large scale screening of the general population establishes the prevalence and incidence of ZES, such guesses should not be taken seriously. A related syndrome, pancreatic cholera, in which severe diarrhea is caused by a hormone other than gastrin produced by an islet tumor is also discussed.--E. S.

21206. FERENCZY, ALEX* and RALPH M. RICHART. (Int. Inst. Study Hum. Reprod., Dep. Pathol., Coll. Physicians Surg., Columbia Univ., New Y... ...nning electron mic...

FIGURE 9: *Examples of informative abstracts from* Biological abstracts.

of it, all chemists, all biologists, etc. But in specialised industrial or research situations where an information scientist as part of a research team is providing a specialised abstracting function for research scientists working in a specific area and where the circulation of the abstract bulletin can be easily controlled, evaluation and criticism may take place. The information scientist here will endeavour to render his colleagues more productive in their research by channelling their reading effort towards the most significant papers. He is able to do this, of course, only if he is well-informed in the subject field and intimately aware of the information requirements of his colleagues.

The *author abstract* is usually informative and is by definition written by the author of the paper abstracted. Author abstracts appear now in most of the primary journals, many of which insist that the author must provide an abstract of any paper he submits for publication. The occurrence of author abstracts has increased enormously in recent years as the editors of scientific journals have begun to accept more responsibility for the ultimate retrieval of the information they generate and have thus insisted that their authors choose meaningful titles for their papers and also provide abstracts of the papers. Author abstracts facilitate the transfer of information in that they can be 'lifted' from the primary journal in their original or in a modified form to be included in abstracting journals. This obviates the need to produce an abstract for every article covered by an abstracting service. Some doubts have been expressed on the validity of author abstracts; it is often maintained that the author is not always sufficiently objective about his paper to produce a well balanced abstract. It should be remembered, however, that the author abstract will invariably be edited before it appears in the pages of a primary journal as part of the normal editorial functions of that journal.

The phrase *telegraphic abstract* is currently in use to refer to a form of computer-generated abstract—but this is really a misnomer as it is not an abstract at all, more precisely it is a set of keywords indicative of the subject content of a paper. The search products of some of the computer-based information retrieval systems such as MEDLARS include for each bibliographical reference that has been output to fulfil the search strategy a citation of indexing terms assigned to the paper by the subject indexers in the creation of the data-base. These terms, which may extend to twenty or thirty, according to the depth of indexing, when considered as a group are tantamount to an indicative abstract.

The phrase *auto-abstract* is also misleading as it refers to a computer produced extraction from a document. The main research into the production of these 'abstracts' was undertaken by H P Luhn at IBM in the late 1950s.[2] Selected sentences were taken from a text on the basis of word frequency and the occurrence of a group of high-frequency words in one sentence. The text of a paper was translated into machine-readable form and then processed by the machine to identify those words which were used with the highest frequency, ignoring uninformative words for which a stop-list was prepared. It was argued that if a word was used with high frequency then it was a significant word and therefore indicative of the subject content of the paper. The machine was then programmed to identify those sentences which included a number of high frequency words—these were output as it was claimed that they were key sentences in which the essence of the paper was concentrated and meaning highly distilled. Obviously such extracts differ basically from the conventionally produced abstract and at their best they could only be indicative of part of the content of a paper. It should again be remembered that the production of auto-abstracts was experimental and that no abstracting service has used these methods in the production of a journal.

The opening definition of abstract included the phrase 'an adequate bibliographical citation to enable the document to be traced'. As the abstract is a secondary source this citation is of vital importance in that it acts as the link between the summary and the original paper—the citation provides the abstract user with the means of accessing the original. There is, as the following examples will show, much variance in the style of citation adopted by the various abstracting and indexing services but all will provide the following essential elements in a citation:

1 The title of the original article; this should preferably precede the author statement as it is more indicative of subject content than a personal name. In some services, *Engineering index,* for instance, the title is for emphasis given in capitalised and emboldened typography preceding the body of the abstract from which it is separated by the remainder of the reference.

2 The author(s) of the paper abstracted. Most abstracting services will endeavour to give the author's place of work or research as this is valued by users of abstracts. It enables them to contact the author for offprints of his paper or to engage in correspondence concerning points covered in the paper. It may also provide enlightening information on the level of the project if the institution in which the author is working is a known centre of excellence.

3 The bibliographical citation. This can consist of a number of elements all of which should be provided by library users when requesting loan copies or photocopies of papers cited in abstract journals. Inaccurate or incomplete bibliographical references can cause much delay in the processing of loans as they will necessitate the checking of the bibliographical details by the lending library. The elements of the citation are:

a) Title of the periodical; this may be abbreviated;
b) Volume number;
c) Part number;
d) Date of issue;
e) Pages over which the article extends; these should always be quoted in full, *eg* 709-11, 717-19, *not* 709-19 or 709 as the latter citations may result in the provision of pages not required which contain extraneous advertising material or the omission of pages which *are* required.

The abbreviations of journal titles used by abstracting services are now usually made according to official national or international standards. Perhaps the first internationally accepted standard, although it was unofficial, was the form of abbreviations used for the periodicals listed in the *World list of scientific periodicals.* The acceptance of the need for standardisation in the field of documentation by official standardising bodies led these organisations to develop standards for voluntary adoption within their countries. There is now a high degree of uniformity between the British and American standards for the abbreviation of titles of periodicals, BS 4148(1970) and ANSI Z39.5(1969) and the International standard ISO 4(1972) is generally in harmony with these documents.

Examples of bibliographical citations used in abstracting journals:

Induction of IgG by lipid A in the newborn mouse. Kolb C, et al. J Exp Med 139; 407-13 1 Feb 74.
Index Medicus

ON THE OXIDATION OF FUEL NITROGEN IN A DIFFUSION FLAME. Sternling, C. V. Shell Dev. Co., Houston, Tex; *J Fuel Soc Jap* v52 n558 Oct 1973 p778-784.
Engineering index

Studies on spermatogenesis in rats. Go V. L. W., Vernon R. G. and Frit. I. B. Banting Best Dept. Med. Res., Univ. Toronto. Canad. J. Biochem. 1971 49/7 (768/775).
Excerpta medica

Structural factors and partition coefficients of pure molten metals. Bergdoll, Meilin S. Univ. Wisconsin Madison Wis. Clin. Toxicol. 000072 5 4 441-51.
CA Condensates

APPLICATIONS OF THIN FILMS IN MICROELECTRICS. J. C. Anderson (Imperial Coll., London, England). *Thin Solid Films (Switzerland)* vol. 12 no. 1, p1-15 (Sep. 1972). (18 refs).
INSPEC

KREKLING (S.) (Psychol. inst., univ. Trondheim, 7000 Trondheim, Norway). SOME ASPECTS OF THE PULRICH EFFECT. Scand. J. Psychol., Swed., (1973), 14, no 2, 87-90, bibl. 21 ref.).
Bulletin signatetique

The CODEN system is now being used as a machine-readable

key to bibliographical citations in abstracts.[3] The system has been developed by the American Society for Testing and Materials (ASTM) and the CODEN register is maintained and updated by the Franklin Institute on behalf of ASTM. CODEN greatly reduce the keyboarding necessary to record the bibliographic identification of each paper and this form of citation greatly reduces the size of the stored bibliographical record. A CODEN is a five character code designating the title of a serial containing four mnemonic characters with a fifth, either A, B or C added to expand the number of CODEN combinations as it is possible that the first four letters of two serial CODEN may be identical, *eg*:

JACP-A *Journal of the American Academy of Child Psychiatry*, and

JACP-B *Jahrbuch fuer Chemiker und Physiker*.

To assist in using the code in computer-based systems an optional check character may also be included—this additional facility is in use by several abstracting services including *Chemical abstracts*. The CODEN for *Information storage and retrieval* is IFSRAS, the check digit 's' is in lower case and is calculated in such a way that common typing errors are detectable by the computer. Each serial has only one CODEN, a new code is designated if a title change takes place.

An International Serials Data System (ISDS) has been established within the framework of the UNESCO UNISIST programme to develop and maintain a register of serials from all countries in all subjects containing all the necessary information for the identification of a serial.[4] The ISDS data files, held at ISDS International Centre, 58, rue de Richelieu, Paris, 2, are maintained in a MARC-like format containing sets of data for the definitive description of each serial. The information will be made available to libraries, secondary information services and individual users. The elements within the ISDS format are:

1 Date of entry or most recent amendment

2 Centre code (national ISDS centres will be established in cooperating countries)

3 ISSN
4 Coden
5 Publication status (current, discontinued, unknown)
6 Type of publication
7 Start date
8 End date
9 Frequency
10 Country of publication
11 Alphabet of original title
12 Language of publication
13 UDC, DC or LC classification
14 Key title
15 Abbreviated title
16 Variant title
17 Former title
18 Succession title
19 Other language edition of
20 Has other language edition
21 Inset in or supplement to
22 Has inset or supplement
23 Related title
24 Imprint
25 Coverage by abstracting services.

The ISSN number (3) is a seven digit number plus a check digit in the form ISSN xxxx-xxxx. This number is not indicative of the information content of the title; it is used solely to identify and to be inseparably associated with a serial title.

Abstracting, indexing and other forms of secondary publications should be seen as the catalogues of the scientific literature: they are the guides which lead users of libraries to the original papers. Self-sufficiency in stock provision was an aim which well funded libraries were able to pursue a century ago before the massive expansion of serial literature had commenced; the extent of the literature is now such that no individual library can ever be self-sufficient. Librarians now aim at providing comprehensive collections of abstracting services rather than attempt to chase the impossible goal of complete in-house serial provision. Abstracting services then enable the library user to gain access to the total stocks of serials available to him via national lending services and networks of library cooperation.

Secondary publications help scientists to maximise the use of the limited amount of time they have available for reading. It has been repeated *ad nauseam* that if a scientist were able to read for twenty-four hours per day for seven days each week he would even then not be able to cover all the literature of potential interest to him. This nevertheless is true, but it assumes that this literature of interest to an individual could be made available conveniently in one collection—this is an idealised view which is rendered unrealistic when one considers the effect of 'scattering'. The papers relating to any given subject will be distributed over a wide span of journals, many of which will have but a tenuous link with that subject. But if our scientist spends his available reading time scanning the 'core' journals in his field —those titles which because of their publishing tradition attract the best papers, and a general scientific or engineering journal, *Nature, Science, Engineering, New scientist,* etc to keep himself informed of developments in science and technology at large, he can use the abstracting services covering his field to identify other papers published in less familiar journals which will be of relevance to him. He is thus putting all the literature through a sieve: this is the current awareness use of abstracting services, their scanning immediately after publication to identify significant papers. The systematic arrangement of the contents of most abstracting journals bringing their abstracts together under specific subject grouping facilitates this scanning. *Chemical abstracts,* for example, is available in five separate groupings— biochemistry, organic chemistry, macromolecular chemistry, applied chemistry and chemical engineering, and physical and analytical chemistry, comprising of eighty sections in all, each group containing a set of sections related by subject matter.

Retrospective search is another major use of abstracting services; in this mode of use the retrospective accumulation of articles in the journals covered by the abstract journal is accessed not by using the main systematically arranged section of the journal but by using the specific subject indexes and author index appended to most abstract journals. The full value of abstracting services in this task should be appreciated—in the absence of these services the searcher would be faced with the daunting task of scanning the numerous individual annual indexes of the periodicals covering the subject. In this retrospective role abstracting services are one of the invaluable research tools of science, providing access to the scientific archive of the primary journals. Various forms of retrospective use can be identified:

a the location, in answer to a subject enquiry, of specific information, *eg* details of a chemical process or the provision of physico-chemical data;

b the compilation of a state of the art review of all the available literature on a topic, the form of exhaustive search made on behalf of an organisation before it decides to embark on an expensive research programme, or by a research student before he chooses his research project;

c the compilation, via the author indexes, of a bibliography of the papers produced by an individual scientist.

Almost fifty percent of the world's scientific and technical literature is published in languages other than English, therefore any scientist whose only reading language is English will be denied access to a substantial proportion of this literature. There is in fact not one 'language barrier' but a series of barriers of varying severity according to the language and the subject field, for example, the majority of English speaking scientists can cope with the French language, and a large number with German, but only a small minority can yet handle Russian and the other Eastern European languages and when the Japanese and Chinese languages are confronted the barrier is virtually complete. All major abstracting services will aim at world coverage of their disciplines or missions and by scanning these as they are published the user can keep himself informed of the major developments in foreign science and technology. Furthermore, the abstract can be the basis on which the decision is made whether to acquire the original. If the original cannot be readily understood because of language difficulties a translation can be acquired, if this does not exist one can be commissioned from a translation centre or a commercial translator.

Abstracting services are frequently used by librarians or information scientists to trace the precise bibliographical reference of a paper for which a library user can remember only partial details. A medical researcher may recall that an author whose surname was 'Rowland' published a paper on 'microencapsulation of steroids' within the last five years but he may not have further bibliographical details. A consultation of the author indexes of *Index medicus* over the last five years will quickly reveal the complete reference. Similarly, a chemist may wish to obtain a paper he has read in a journal sometime during the last three years on 'the extraction of gold from sea water' but whose reference he has mislaid, the only key he may have is the title of the paper and the subject matter. Again this paper could be easily located by reference to the subject index of *Metals abstracts*. If the abstract journal was not available to assist him in this search the library user would once more be reduced to scanning through the successive annual indexes, if these existed, to a wide range of 'likely' journals—assuming, of course, that he had access to these journals.

REFERENCES

1 Collinson, R L *Abstracts and abstracting services* ABC Clio Inc, Santa Barbara, California; Oxford, England, 1971, p 11.

2 Luhn, H P 'The automatic creation of literature abstracts' *IBM journal of research and development* 2 (2), 1958, pp 159-65.

3 Batik, A L 'The Coden system' *Journal of chemical documentation* 13 (3), 1973, pp 111-3.

4 Rosenbaum, M 'International serials data system' *International cataloguing,* Oct/Dec 1972, pp 4-6.

7
Access to the journal literature

3: Abstracting services, their development, some problems and solutions, and the impact of the computer

ABSTRACTS CAN BE TRACED back to the very beginnings of the periodical literature of science and technology; most of the early journals were in part abstracting organs which aimed at summarising the significant papers appearing in the contemporary literature. It has already been emphasised that the early learned journals were catholic in coverage embracing contributions from the arts, the sciences, theology and the legal profession. The first abstracting publications reflected the nature of the journals on which they reported—it was only with the emergence of the literature of chemistry and physics in the late eighteenth century that the specialised abstracting journal as we know it today appeared. Virtually all of the early abstracting publications resulted from the heroic efforts of energetic individuals who sought to serve scholarship by providing the means for their 'philosopher' colleagues to become aware of the efforts of their peers at home and abroad. With the development of the specialised literature the task of reading, assimilating and summarising the contents of many thousands of communications moved beyond the powers of even the most selfless abstractor-scholar. The task of controlling the growing tide of information was thus assumed by the professional and learned societies who saw this organisation and dissemination process as one of their basic duties to the members they served. Consequently they were willing to financially subsidise the publication of abstracting journals—organs which they saw as the basic tools of research. Abstracting services could never have flourished without this underpinning from the societies, this aspect of publishing was not attractive to the commercial houses because of the high cost of labour involved in the production of acceptable abstracts and the relatively small circulation of the resulting journals.

The first specialised abstracting journal, *Chemisches journal fur die freunde der naturlehre,* founded by Lorenz von Crell who has been called the founder of chemical journalism, was issued irregularly in six volumes between 1778 and 1781. The journal's reputation was such that an English translation was published in three volumes in 1791 and 1793 as *Crell's chemical journal, giving an account of the latest discoveries in chemistry.* The *Analytical review* published in London under the editorship of Joseph Johnson flourished for ten years from 1788; this was, as its subtitle will indicate, similar in character and coverage to some of the earliest learned journals: 'history of literature domestic and foreign, scientific abstracts of important works in English, articles or reviews of foreign books, collections of new pieces of music and works of art and the literary intelligence of Europe'.

The most influential of the earlier abstracting journals was the weekly *Pharmaceutisches central-blatt* published by the Berlin Academy between 1830 and 1849 under the editorship of Gustav Theodor Fechner. This publication was to become a model for many of the subsequent abstracting services, its influence on the nineteenth century services was similar to that of *Chemical abstracts* for the following century. As the scope of this journal extended it was in 1830 retitled *Chemisches-pharmaceutisches central-blatt* and in 1856 the journal adopted the now familiar name *Chemisches zentralblatt* which it retained until its demise in 1970. *Die fortschritte der physik* founded in 1845 by the Deutsche Physikalische Gesellschaft to control the rapidly expanding literature of physics was another of the classical services which in 1920 assumed the title *Physikalische berichte*.

The English language abstracting services started to emerge towards the end of the nineteenth century. Abstracts were first published in the *Journal of the chemical society* in 1871, from 1923 these abstracts were issued in a separate publication by the Bureau of Abstracts as *Abstracts of chemical papers,* the service ultimately became *British abstracts* which ceased publication in 1953 as it was then merely duplicating the coverage of more adequately funded American services. One of the major British services was founded in 1895 when the Physical Society, London, commenced publication of *Abstracts of physical papers from foreign sources,* a journal which was continued in 1898 by the Institution of Electrical Engineers as *Physics abstracts,* series A of *Science abstracts.*

In the United States the major engineering abstracting service *Engineering index* was published by the Association of Engineering Societies from 1884. The following decade saw the institution of the most celebrated of all abstracting services, *Chemical abstracts.* The Massachusetts Institute of Technology published *Reviews of American chemistry* from 1895 to 1906—this was continued in 1907 by the American Chemical Society as *Chemical abstracts.*

The major abstracting services serving the established scientific disciplines had been established by the early twentieth century, their essential features were usually common—the frequency of publication was monthly, the arrangement was under broad subject according to a classification and this was supplemented by specific subject and author indexes to aid specific search. But this pattern is now evolving rapidly towards a more flexible organisation determined by changing information needs.

There are now approximately 2,000 abstracting and indexing services covering science and technology, these are listed in a number of useful bibliographies:

National Federation of Science Abstracting and Indexing Services *A guide to the world's abstracting and indexing services in science and technology.* Lists 1,855 services arranged alphabetically under their titles. Information is given on date of institution, frequency, average number of abstracts per year, subject coverage, method of arrangement and price. The main sequence is supplemented by a UDC classified index and country and subject title listings.

International Federation for Documentation. *Abstracting services Vol 1: Science, technology, medicine, agriculture* 2nd ed The Hague, 1969. Provides information on 1,300 separately published services. Arrangement is alphabetical by title and this is supplemented by a country index and a UDC title listing with subject index. It should be noted that this bibliography is limited to abstracting services, indexing and contents listing services are excluded. Information given for each service includes the number of abstracts published each year, length of abstract and the number of journals monitored by the service. The most recent bibliography of English language secondary services is the National Lending Library (now British Library, Lending Division) *KWIC index to English language abstracting and indexing publications currently being received by the NLL* 4th ed Boston Spa, 1972. Approximately 1,300 titles are listed alphabetically under the subject keywords they contain. No bibliographical information is given.

Ulrich's international periodicals directory 14th ed 1971/72 lists some 1,300 separately published secondary services under the heading 'abstracting and indexing services'—this has been a feature of all recently published editions of *Ulrich*.

One of the main problems facing the publishers of abstracting services is that of currency—keeping the time-lag between the publication of the paper in the primary journal and the appearance of the abstract in the secondary publication to a minimum. Excessive delays in the process of abstracting detracts considerably from the current awareness use of abstract journals and this alerting use which is made of secondary sources represents their major role. Before computer-aided publication, delays of from six to nine months were commonplace, but in recent years some of the major services have by the application of computers been able to reduce this period to between two to four months. The National Federation of Scientific Abstracting and Indexing Services (NFSAIS) was established in the United States in 1958 as a forum for cooperation between the fourteen major services and one of its chief aims was to apply new technology in the preparation of abstract journals to expedite their currency. Virtually all of its member services are now using computer-aided preparation. Another device which has been used to combat delay has been the provision of page proofs or advance copies of primary material via air mail to the secondary services. This procedure was stimulated mainly by the Abstracting Board of the International

Council of Scientific Unions whose remit is to assist its member journals to improve their speed, quality and coverage. The Board sustains American, British, French and Russian cooperation within physics, chemistry, botany and astronomy, its member journals being *Chemical abstracts, Physics abstracts, Physikalisch berichte, Bulletin signaletique, Referatifnyi zhurnal, Biological abstracts* and *Astronomischer jahresbuch*—it should be remembered that *Bulletin signaletique* and *Referatifnyi zhurnal* cover all four disciplines.

Indexing services such as *Applied science and technology index* and *British technology index* (BTI) have existed for many years and one of their functions has been to publish their entries with the minimum of delay, thus acting as current awareness tools. They provide less information than the abstracting services about the subject content of the papers they cover in that they simply arrange their bibliographical citations to articles under a series of alphabetical subject headings. BTI endeavours, with a high degree of success, to publish its entries with an average delay of one month but with ASTI considerably longer periods of up to six months are often experienced.

An increasingly popular form of current awareness aid is the contents list publication which contains reproductions of journal contents lists very shortly after the originals are published. The best known of this class are the Institute for Scientific Information's (ISI) *Current contents* series of six publications covering the chemical and physical, life, agricultural, biology and environmental, and social and behavioural sciences. Each section is published weekly and in some instances the contents notification is in print before the journal it covers as ISI has secured pre-publication rights from many of the journals it monitors. The six sections of *Current contents* now cover more than 5,000 primary journals. Other aids to currency are the SDI services using machine-readable data bases (see page 93), *CA Condensates* and BioSciences Information Service (BIOSIS) for instance provide output some weeks in advance of the corresponding abstract journal publications.

There are approximately 2,000 individual abstracting and indexing services covering science and technology throughout the world. This proliferation results in a considerable amount of duplication and overlapping: two or more services will be abstracting the same article. There is also a considerable proportion of the literature which is not abstracted, the term 'underlap' has been used to denote the non-coverage of journals by secondary services. Perhaps the most extensive examination of this duplication and omission was undertaken by Martyn and Slater in the Aslib Research Department where some twenty highly specific and presumably exhaustive bibliographies were examined in an attempt to ascertain what proportion of their citations were traceable by using the appropriate abstracting and indexing services.[1,2] The twenty bibliographies produced a total of 3,420 separate citations—of this number twenty one percent had not been covered by any service, thirty two percent had been covered once and forty seven percent more than once. The citations which had not been picked up by the secondary services were scrutinised to discover the reason for their omission but no common factor could be identified and it was concluded that there was no apparent reason for their non-coverage: the papers were not of inferior quality to those covered, there was no language bias in those omitted, etc. What emerged from the study was the value of a good bibliography, by using the most appropriate abstracting journals to search on a topic it would appear that the user could expect to capture about eighty percent of the existing relevant papers and this figure could be increased to near ninety percent by using other relevant abstracting services, but the searcher wanting total coverage of his subject would not achieve it by using abstracting services alone. These studies were made in the mid nineteen sixties and the coverage of services has no doubt improved since then but the inherent structure of the periodical literature and the scatter phenomenon make Martyn's findings still tenable.

Overlapping of journals is wasteful in that the effort of abstracting is being duplicated or triplicated. There are, no doubt, instances where the same paper must be abstracted for different readerships and then slanting rather than duplication will occur but a more rationally concerted approach to coverage between services is needed to minimise straightforward duplication. The combined effect of omission and overlapping on the users of abstracting services is that they can never be sure that they are monitoring all the literature by using abstracts and that even when they subscribe to a number of services they must tolerate a degree of duplication and a degree of total underlap.

A cooperative approach to the problem is being made by Chemical Abstracts Service (CAS), BioSciences Information Service of Biological Abstracts (BIOSIS) and Engineering Index, Inc who in 1970 launched a multiphased study of the coverage of their journals. Phase one[3] and two,[4] the overlap between the lists of

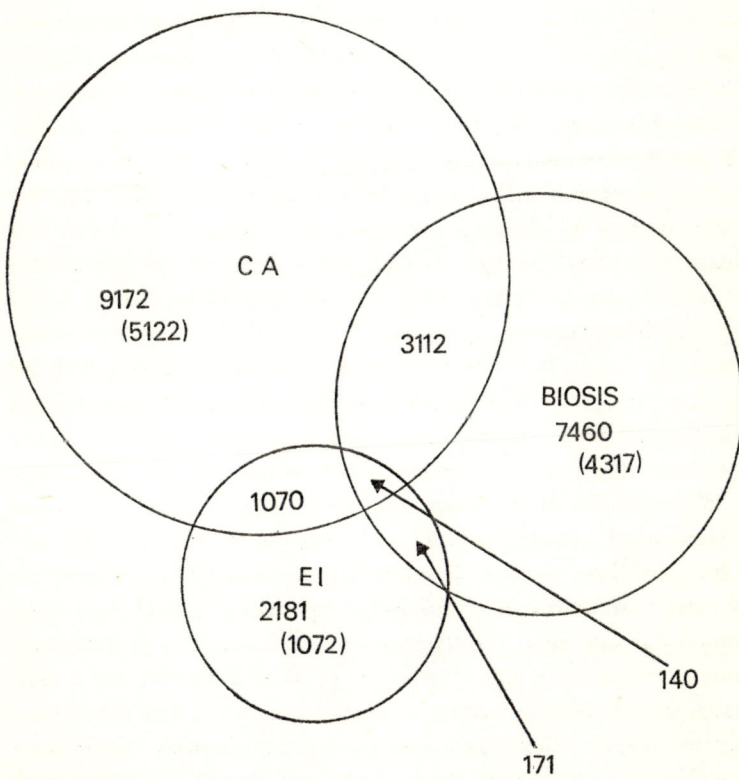

FIGURE 10: *Overlap in the lists of journals monitored by* Biosciences Information Service (BIOSIS), Chemical Abstracts Service (CAS) *and* Engineering Index (EI).

journals monitored by the three services and the overlap among the journal articles selected for coverage, have been completed. The first study indicated that as of May 1970 BIOSIS, CAS and EI were collectively covering 14,592 different titles; 10,511 journals, seventy two percent of the total were covered by one service only, 3,948, twenty seven percent, by two of the services, and 140 journals, one percent, by all three. CAS monitored 5,122 journals uniquely, BIOSIS 4,317 and EI 1,072. The critical two way overlaps were between BIOSIS and CAS and CAS and EI: BIOSIS and CAS contained entries from 13,060 different currently published journals and of these 3,112 or 23.8 percent were covered by both; CAS and EI covered 10,815 different currently published journals of which 1,078 or ten percent were covered jointly. The overlap between BIOSIS and EI was insignificant, of the 9,470 titles covered by these services only 171 titles or 1.8 percent were commonly covered. The second phase of the project was to examine overlap at the level of individual article in the titles identified in phase one as being covered by more than one service. For the 140 journals covered by all three services it was demonstrated that the maximum possible article overlap was an insignificant 822 articles from the total of 29,182 articles which might have been covered by all three services. Again the two way overlap between BIOSIS and CAS and CAS and EI gave cause for concern, BIOSIS and CAS overlapped 48,856 articles and CAS and EI 17,484. In the face of this evidence further investigations are being undertaken as a preliminary step to rationalisation of coverage.

Martyn and Slater examined the possible causes of loss of information between primary publications and abstraction and identified three sources. The first factor, non-coverage of primary journals by abstracting and indexing services has already been noted but the others, faulty indexing by abstracting journals and non-availability of primary journals when required by a user, need further consideration. No detailed evaluation of the quality of the subject indexes provided by abstract journals has yet been carried out but regular users of abstract journals will be aware of inconsistencies which will be encountered in virtually every index. This problem is further aggravated by the increasing use of keyword-title indexes; some abstract journals now rely solely on the uncontrolled title keywords as the method of subject access to their abstracts. Martyn estimates that the user of an abstract journal will find no more than seventy five percent of the potential eighty percent coverage which can usually be obtained by using an abstract journal and that he is unlikely to find more than fifty percent of the material unless he possesses a good knowledge of the subject or exercises considerable resourcefulness while using the subject index. 'The users of abstracts should be gifted with a great deal of ingenuity or, as a substitute for this, a reasonably wide knowledge of the field so that all possible headings under which material may be discovered can be inspected.'

The non-availability of the material identified through an abstract journal should not now be a major problem, but a constraint which may hinder the users of abstract journals is their ignorance of existing back-up services such as lending, photocopying and translation agencies. If users are not made aware that these services exist their usage of secondary services will be seriously inhibited: they may believe that the only material which can be obtained readily is the literature available in their local libraries. The existence of the British Library Lending Division (BLLD), formerly the National Lending Library for Science and Technology, transforms the abstracting and indexing services into that library's catalogue. In addition to fulfilling its major function of supplying the world's scientific and technical literature to users in the United Kingdom the BLLD is now used extensively by libraries overseas. One of the library's largest growth areas in recent years has been its overseas photocopying service. Although such a single highly comprehensive loan collection is not available in the United States or elsewhere literature users in these countries are usually well served by their academic, industrial and, in the cases of large centres of population, public technical libraries. If these sources cannot produce the required papers many of the abstracting services themselves undertake to supply their users with copies of the papers they cover, *eg Chemical abstracts* will provide a copy of any Russian paper it covers and the Institute for Scientific Information through its OATS (Original article tearsheet) service will provide the actual original, extracted from a copy of the journal, of any paper listed in its publications.

Adams and Baker have suggested that the organisation of abstracting services reflects the organisation of the sciences, they maintain that with the large scale public funding of technology resulting in mission oriented projects, scientific institutions are being subjected to fundamental changes.[5] New information services are being developed to meet the requirements of the centrally funded, mission oriented projects and these are in conflict with traditional aims of the old established discipline oriented information services. Discipline is defined as 'a body of knowledge empirically organised for purposes of transmission through teaching'. A discipline in science is an academically transmitted corpus of knowledge and the discipline oriented service reports its accretions. The world's major abstracting services all developed within the disciplinary tradition. Mission oriented research is characterised by its associated with the solution of a problem or a set of problems relating to publicly identified social goals such as the eradication of the biodeterioration of materials or the conquest of cancer. The information services needed to feed these missions cut across the previously accepted disciplinary boundaries and they cannot use as their base a traditionally organised single disciplinary service, they must draw and collate their information from more than one discipline. Adams and Baker summarise the problem thus: 'discipline oriented science presents classic, relatively constant and continuous forms and requirements; mission oriented science encourages the new organisational forms of unproven stability and indefinite duration. This conversion of scientific institutions from the former base to the latter is a dynamic and stressful feature of the twentieth century scientific revolution.'

The former static and monolithic nature of the major abstracting services published in a rigid, single format could not: a) effectively handle the increasing mass of knowledge, or b) attempt to cater for the shifting needs of mission oriented projects. The application of computer processing methods has, however, given the disciplinary based services the increased flexibility necessary to adapt to the changing patterns of research and to repackage their contents to the requirements of groups and individuals. In a computer based abstracting system information can be culled from the source documents in a single abstracting/indexing operation and input to the system in a single keyboarding operation. Information can be manipulated, assembled and packaged in a variety of contexts from the resulting machine readable data base.

The computer was first applied to the production of an indexing publication in 1961 when the Chemical Abstracts Service (CAS) introduced the KWIC (Keyword-in-context) index (see page 98) *Chemical titles* as a current awareness aid to highlight the papers published in 600 of the most prestigious titles covered by the parent *Chemical abstracts,* which was then still being published in its traditional format. By the end of the 1960s a whole range of new CAS publications, the CAS Section Groupings, *Chemical-biological activities, Polymer science and technology,* etc, had been produced from a machine readable data base derived from a single intellectual analysis of the 13,000 source journals covered by CAS, to serve the interdisciplinary needs of scientists. Computerisation was steadily adopted by other services. The computer based Medical Literature Analysis and Retrieval System MEDLARS, introduced initially to expedite the publication of *Index medicus,* whose publication schedules were fighting a losing battle with the rapid expansion of the medical literature, became operational at the national library of Medicine, Washington, in 1964. In the same year Biological Abstracts Inc assumed the title Biological Information Service of Biological Abstracts to reflect its broadening scope and the diversification of information products accessible from its computer data base. In the United Kingdom the Institution of Electrical Engineers, publishers of *Science abstracts* (Part A: *Physics abstracts* and B: *Electrical engineering abstracts*) since 1898, had by 1969 converted its abstracting and indexing operations into the fully integrated computer-based system INSPEC (Information Services in Physics, Electronics and Control). The end of the decade saw the complete reorganisation of the major discipline-oriented abstracting services with their adoption of computer-based principles.

Computerisation became inevitable with the proliferation of the literature; abstracting services could no longer rely on manual methods of sorting and preparation of data for conventional hot-metal typesetting. The volume of the scientific literature increased more than fourfold between 1957 and 1970 with an annual growth

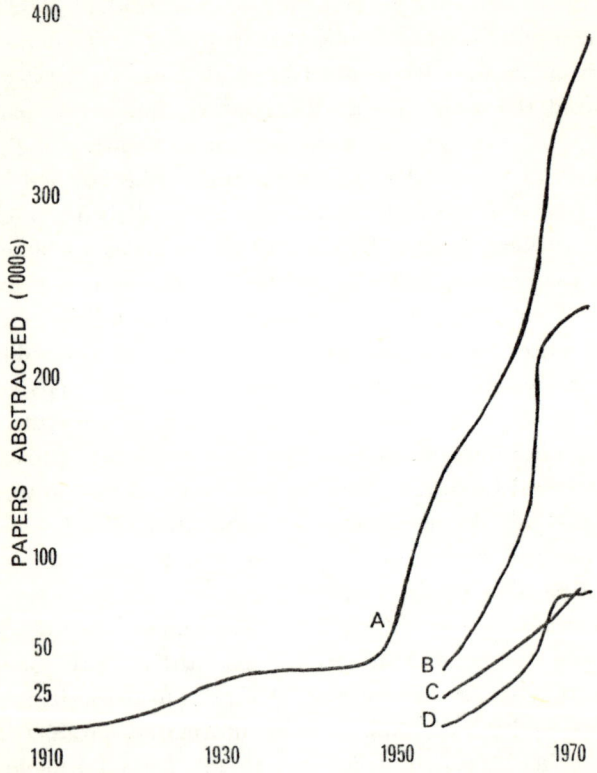

FIGURE 11: *Expansion in the numbers of papers abstracted by* Chemical abstracts, Biological abstracts, Engineering index *and* Physics abstracts. A=*Chemical abstracts;* B=*Biological abstracts;* C=*Engineering index;* D=*Physics abstracts.*

rate of 10.6 percent.[6] The increase in the number of items abstracted each year for *Chemical abstracts* is illustrated in figure 11, this growth has been exponential since the 1950s—in 1957 100,000 items were abstracted but by 1974 the number was almost 400,000.

This pattern is common to other services, *Biological abstracts* published 40,000 abstracts in 1957 and 240,00 in 1974, *Engineering index*, 26,000 and 70,000 in the corresponding years. This ever increasing volume of material resulted in longer and longer delays between the receipt of the original paper and the appearance of the abstract in a secondary journal. Computerisation offered the only means of cutting back on this delay or at least of arresting it.

The emergence of new specialisations and the development of interdisciplinary research resulted in the requirements of scientists and technologists being more selective and these could not be adequately satisfied by services which simply presented their abstracts in a conventionally published arrangement. Scanning such publications, which were ever growing in size, became increasingly time consuming as the number of abstracts to be rejected within each intendedly specific section grew accordingly. The personalised current awareness service offered by a computer based system was seen as a solution to this problem. Conventional presentation of the abstracting journal was not superseded—this would always be necessary for retrospective access to the literature in libraries but the machine gave abstracting services the facility for dual packaging; conventionally formatted abstract journals to cater for library and archival use and personalised selective dissemination of information (SDI) services designed for mission oriented groups and individuals with specialised requirements.

The computer based SDI services were developed by the mid 1960s. In an SDI environment a profile comprising of keywords indicative of information needs is constructed for each user of the system. The profiles are committed to magnetic tapes and these are matched each week against the service's data base: where there is coincidence between the keywords in a user profile and the keywords assigned to a document in the data base, either by the original author as part of the title of that document or by an indexer of the system using a controlled indexing language, the document is considered a 'hit' and is notified to the user as being relevant to him in his weekly card or standard line printer output. The United Kingdom Chemical Information Service

(UKCIS) at Nottingham University introduced the Chemical-Biological Activities (CBAC) service, based on a section of the CAS data base in 1966 and by 1969 subscribers could avail themselves of the fully comprehensive *CA Condensates* services covering the total CA data base of 13,000 journals, plus reports and patent specifications, which brought citations to the attention of the user on average ten weeks after their publication in the primary journals. The Biosciences Information Service of Biological Abstracts

```
UPDATE OF  PROFILE 0501        16/10/73
       PROFILE RECORD
PROFILE LENGTH          93    (MAX=1020)
PROFILE WORKING

00    1    0501-X

93    1    COGNITION PROCESSES, ETC

98    1    ABC

91    1    71 OR 72 OR 73 OR 74 OR 75

71    1    01 AND NOT 02 / COGNITION, RECOGNITION AND PERCEPTION
      2    PROCESSES, INCLUDING PATTERN RECOGNITION BUT NOT
      3    CHARACTER RECOGNITION. AUTOMATA

72    1    03 AND 04 / LEARNING OR ADAPTIVE MACHINES AND SYSTEMS

73    1    04 AND 06 / MACHINE INTELLIGENCE

74    1    05 AND (06 OR 07) / MODELS OF INTELLIGENCE, BRAINS OR
      2    NEURAL SYSTEMS

75    1    03 AND 08 / ANYTHING BY THE TWO AUTHORS ON LEARNING OR
      2    ADAPTIVE SYSTEMS

01    1    *COGNIT*
      2    PERCEPT*
      3    ROBOT*
      4    AUTOMATA
      5    AUTOMATON

02    1    CHARACTER*

03    1    LEARNING
```

FIGURE 12: *An example of an* INSPEC SDI *profile.*

(BIOSIS) introduced its Current Literature Alerting Search Service (CLASS) providing semi-monthly printouts of citations tailormade to a client's interests five weeks in advance of their appearance in the issues of *Biological abstracts*. The Institution of Electrical

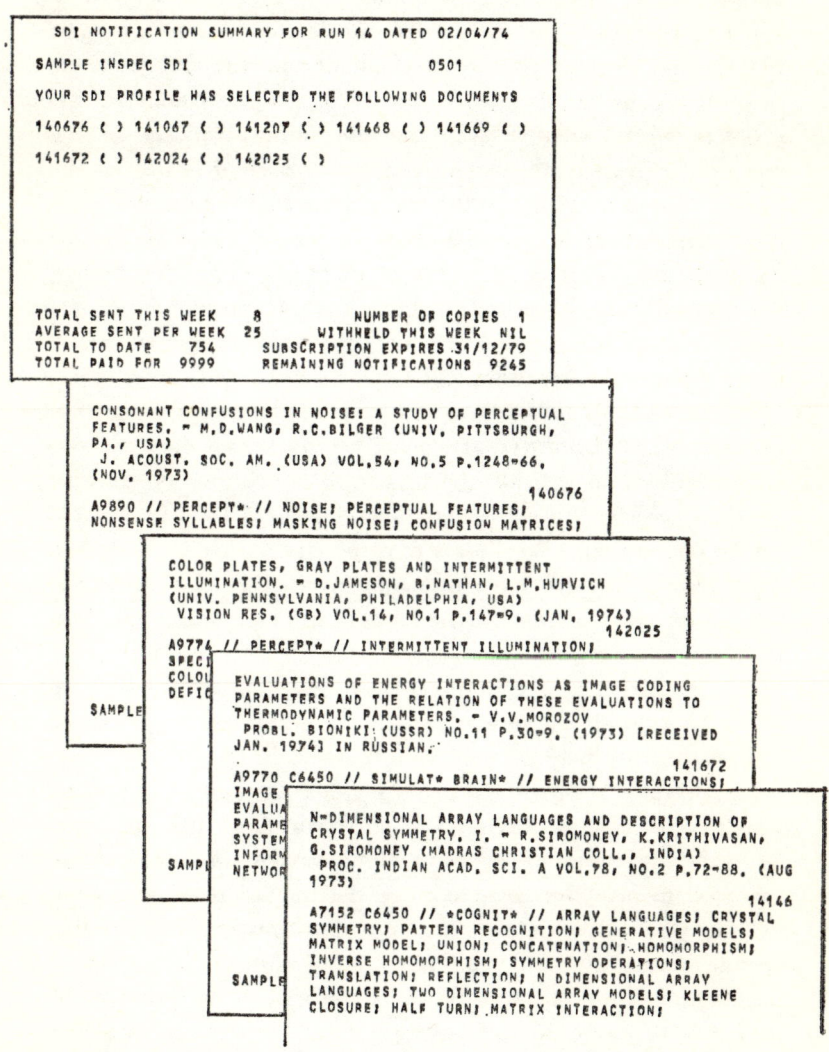

FIGURE 13: *Output references for the profile illustrated in figure 12.*

Engineers introduced INSPEC as a fully operational SDI service in 1970.

The retrospective manual searching of the literature can be time consuming when the back file indexes of abstracting service are consulted under numerous indexing tags for searches in which a number of parameters are interrelated. A search may be required for instance on the 'effects of steroid compounds on the integral organs of laboratory animals'. A generic search on this problem would involve scanning the subject indexes of the abstracting journal under a very large number of terms; all the known steroid compounds, the various internal organs, and each individual laboratory animal—that is, if the search was to be comprehensive. A state of the art search of this dimension carried out manually would indeed be daunting, not to say tedious. If a search strategy is presented to a computer based system as a statement of the indexing terms and their relationship to each other, *eg* STEROIDS (all) *and* INTERNAL ORGANS (all) *and* LABORATORY ANIMALS (all) *but not* RATS, the machine can search the data base retrospectively to provide the user with a printout of relevant citations which satisfy the search strategy. All the above mentioned services offer their users retrospective search facilities, the UKCIS back file service dates from 1962 for the *Chemical titles* base to 1968 for the *CA Condensates* base, BIOSIS offer search facilities back to 1959 and the INSPEC base is accessible from 1969. Full details of forty eight computer based information services are given in *A guide to computer-based information services* compiled by Ruth Finer and published by ASLIB in 1972.

In addition to providing current awareness and retrospective search facilities most services will make copies of their machine-readable data bases available to organisations who wish to use them for the provision of an in-house alerting and retrieval system. For organisations with an extensive research programme this may be more economical than placing a large number of individual profiles for subscription with the service. The current cost of obtaining the INSPEC data bases for physics; electrical engineering and electronics; and computers and control is £2,350 per annum with abstracts and £1,750 without abstracts but with full indexing information; the average cost of an individual profile per annum on one data base and containing up to fifty search terms is £65, for two bases £85 and for three bases £105.

The computer can now give the user of the periodical literature the facility of interrogating the disciplinary data bases on line. It is anticipated that in the very near future all of the major abstracting services will offer on line access to their citations as, for instance, the MEDLARS and INSPEC services do already. Keyboard terminals, available in libraries and information centres, will enable the user to converse with the system in a conversational iterative mode. He will be able to devise a search strategy consisting of a statement of indexing terms expressing his information need, and to test its suitability by assessing the relevance of the references it retrieves. If these are not pertinent the initial search strategy can be modified, broadened or made more specific by subtracting or adding search terms, until the acceptable degree of relevance is attained. On line access is, of course, almost instantaneous and thus eliminates the delays experienced in the receipt of output by the users of existing retrospective services. This is caused by the batch processing of requests where the system will delay the processing of an individual search until sufficient requests have been received, may be thirty or forty, to justify the expense of machine time by simultaneous processing.

The MEDLARS data base can now be used on line through the MEDLINE (MEDLARS on line) system which is accessed via a teletype terminal connected to the National Library of Medicine in Bethesda, Maryland, through a communication linkage provided by an international network using telephone lines. The search process, including a preliminary discussion of search strategy between a librarian and a requester usually takes up to thirty minutes and the average search cost is $7.50. In the United Kingdom INSPEC have introduced on line searching, RETROSPEC 1 is a retrieval system covering the literature on computers and control engineering. The data base consists of English language journal papers selected from the INSPEC computer and control base, and access to the system is via the post office telephone service using a teletype compatible terminal through the bureau service operated

```
ANIONS +INTERACTION OF ALUMINUM CATIONS WITH HYPO PHOSPHATE CCCCAK-0036-3845
N SPECTRA OF THE MOLTEN ALUMINUM CHLORIDE - ZINC CHLORIDE     INUCAF-0008-0079
L METH ACRYLATE - ETHYL ALUMINUM CHLORIDE POLYMERIZATION.=+Y JPYBAN-0010-0035
IVES OF AROMATIC ORGANO ALUMINUM COMPOUNDS.= ALK OXY DERIVAT DANKAS-0201-1363
ECTRIC DETERMINATION OF ALUMINUM CONTECT IN A COATING DURING ZVDLAU-0037-1484A
FECT OF TEMPERATURE AND ALUMINUM CONTENT ON THE DUCTILITY    IVUMAX-14-12-058
F ZINC SULFIDE - COPPER ALUMINUM DI SULFIDE AND ZINC SULFIDE ZPSBAX-0015-1030
AND ZINC SULFIDE SILVER ALUMINUM DI SULFIDE PHOSPHORS.=+IDE  ZPSBAX-0015-1030
LTEN LITHIUM FLUORIDE - ALUMINUM FLUORIDE MIXTURES.=  +OF MO CCCCAK-0036-3746
YSTEM SODIUM FLUORIDE - ALUMINUM FLUORIDE.  ACTIVITIES IN    MTGTBF-0003-0495
YMIUM(3+) DOPED YTTRIUM ALUMINUM GARNET LASER.=+ WITH A NEOD IEJQA7-0008-0023
PROPERTIES OF YTTERBIUM ALUMINUM GARNET SINGLE CRYSTALS.= PR IVUFAC-14-12-088
OMIUM(3+) DOPED YTTRIUM ALUMINUM GARNET.=  +LIFETIME OF CHR  JPSOAW-0004-3258
TY OF ROLLED METAL FROM ALUMINUM GRAINS.=             +HEREDI PMANAI-11-12-025
 REACTIONS WITH LITHIUM ALUMINUM HYDRIDE.=         +AND THEIR CRBRAT-0021-0037
O COMPOUND WITH LITHIUM ALUMINUM HYDRIDE.=+ BI CYCLIC NITROS TELEAY-1972-0007
S IN VACCINES SORBED ON ALUMINUM HYDROXIDE GEL.=  +COMPONENT ZMEIAV-49-01-062
RACTION OF THE HYDROSOL ALUMINUM HYDROXIDE WITH POLY ACRYLIC DANUAO-28-07-039
E+ELECTRODE KINETICS OF ALUMINUM IN CHLORIDE MELTS WITH RESP ELCAAV-0017-0119
TION OF ISO BUTENE WITH ALUMINUM IODIDE IN THE PRESENCE OF   MACEAK-0151-0139
RMODYNAMIC FUNCTIONS OF ALUMINUM ION CHELATES WITH CHROMOTRO RRCHAX-0016-1849
ILITY OF SILICON IN THE ALUMINUM METALLIZATION OF INTEGRATED APPLAB-0019-0524
CHARACTERISTI+EFFECT OF ALUMINUM ON THE BETA TRANSFORMATION  JNUMAM-0042-0227
 IRON(3+) DOPED LITHIUM ALUMINUM OXIDE (1/5/8).=         +IN CHPLBC-0012-0062
ARGE SINGLE CRYSTALS OF ALUMINUM OXIDE BY A GRADIENT FURNACE JCRGAE-0011-0345

PORATION IN DOPED ALPHA ALUMINUM OXIDE BY HIGH TEMPERATURE   CHPLBC-0012-0012
 MOLYBDENUM TRI OXIDE - ALUMINUM OXIDE CATALYST.=    +ON THE NEFTAH-0011-0850
E ABSORPTION SPECTRA OF ALUMINUM OXIDE CRYSTALS, ACTIVATED   ZPSBAX-0015-1008
 SORPTION PROPERTIES OF ALUMINUM OXIDE TREATED WITH BORON    VBSKAK-71-06-015
STAL EQUILIBRIUM IN THE ALUMINUM POTASSIUM ALUM TETRACOSA    IVUTAK-14-05-067
D SCRUBBER FOR CLEANING ALUMINUM PRODUCTION GASES.=    +SPEE TVMTAX-44-12-028
15N9YU CHROMIUM-NICKEL- ALUMINUM STEEL.=       +STRENGTH OF K FKMMAJ-07-05-091
5-DE+RUBIDIUM SULFATE - ALUMINUM SULFATE - WATER SYSTEM AT 2 ZNOKAQ-0017-0256
LUM TETRACOSA HYDRATE - ALUMINUM SULFATE OCTADECA HYDRATE    IVUTAK-14-05-067
ENGTH OF IRON-CHROMIUM- ALUMINUM SYSTEM ALLOYS.=      +AND ST MTOMAX-71-12-053
HASE REGION OF AN IRON- ALUMINUM SYSTEM STUDIED BY POLAROGRA ELKKAX-0007-1884B
FFUSION IN THE NIOBIUM- ALUMINUM SYSTEM.=       KINETICS OF D MTGTBF-0003-0421
OYS IN THE IRON-NICKEL- ALUMINUM SYSTEM.=  +HIGH COERCIVE AL VMUKA5-0012-0736
  ACOUSTIC EMISSION OF ALUMINUM TITANATE.=                   JACTAW-0055-0054
METHYL CYCLO PENTADIENE ALUMINUM TRI CHLORIDE COMPLEXES.=+A  JACSAT-0094-0289
ROWTH AND NUCLEATION OF ALUMINUM TRI HYDROXIDE FROM SEEDED   JCRGAE-0012-0039
IN IN TWISTED 99.7 PCT. ALUMINUM WIRES.=    +MEAN SHEAR STRA APAHAQ-0030-0399
TINIZED AND CHLORINATED ALUMINUM. HYDROGENATION OF BENZENE   CHDCAQ-0273-1704
E BROMINATION OF MOLTEN ALUMINUM.=              KINETICS OF TH ZPKHAB-0044-2740
MIDE IONS ON SILVER AND ALUMINUM.=         + CHLORIDE AND BRO JICSAH-0048-0911
HARDENED AND IRRADIATED ALUMINUM.=          +FRICTION IN WORK MRMTAU-0068-0793
CTROLYTIC PRODUCTION OF ALUMINUM.=      +ANODE DURING THE ELE TVMTAX-44-12-025
YER DEFECTS IN QUENCHED ALUMINUM.=      +OF THREE AND FOUR LA PHMAA4-0025-0449
INTERACTION ON LEAD AND ALUMINUM.=  +ON THE ELECTRON-PHONON  LNUCAE-0002-1228
MOLYBDENUM - 1 WT. PCT. ALUMINUM.=  +ZIRCONIUM - 3 WT. PCT.  JNUMAM-0042-0122
```

```
LECTROPHORESIS IN ACRYL AMIDE GEL IN IMMUNOCHEMICAL STUDIES. LABDAZ-1972-0049
EMOGLOBIN IN POLY ACRYL AMIDE GEL.=     +ELECTROPHORESIS OF H UBZHAZ-0043-0765
S UNDER HYPOTHERMIA.=   AMIDE GROUPS OF GOPHER BRAIN PROTEIN UBZHAZ-0043-0697
OF DIAZO ACETYL GLYCINE AMIDE IN MICE.=  SPERMATIC TOXICITY  PLRCAT-0003-0133
F ALLYL ISO PROPYL ACET AMIDE INDUCED INCREASE OF DELTA-AMIN ABBIA4-0148-0010
N ISOLATED SPI+ACETAZOL AMIDE INHIBITION OF PHOTOSYSTEM II I PYTCAS-0011-0065
 HYDROLYSIS OF SPECIFIC AMIDE SUBSTRATES.=           +CATALYZED JACSAT-0094-0293
N/,N/-TETRA ETHYL THIUR AMIDE SULFIDE IN THEIR EFFECT ON     FATOAO-0035-0106
IVES OF ANILINE IN FORM AMIDE SYSTEMS.=+ AND ALK OXY DERIVAT CANWAJ-0016-1217
A METHYL PHOSPHORIC TRI AMIDE.  COMPARISON WITH P-NITRO BENZ TELEAY-1972-0165
D NON-PLANARITY OF CYAN AMIDE. COMPLETE NEGLECT OF DIFFERENT JMOSB4-0010-0285
TROPY OF PHOSPHORYL TRI AMIDE.=        +HEAT CAPACITY AND EN JCEAAX-0017-0094
ION OF N-METHYL PROPION AMIDE.=       +STUDY OF AUTO ASSOCIAT CHDCAQ-0273-1693
ERIVED FROM PHENYL ACET AMIDE.=    +A,B-UNSATURATED AMIDES D RRCHAX-0016-1925
IN HEXA METHYL PHOSPHOR AMIDE.= +CHLORO SILANE / MAGNESIUM   TELEAY-1972-0075
E RESIN WITH DI CYAN DI AMIDE.=+ THE HARDENING OF AN EP OXID JAPNAB-0015-3067
CONES IN DI METHYL FORM AMIDE.=+HALF WAVE POTENTIALS OF CHAL JOCEAH-0037-0106
6 IN N,N-DI METHYL FORM AMIDE.=+OF SODIUM DI BENZO-18-CROWN- JACSAT-0093-7133
LEX TRIS(DI METHYL FORM AMIDE)IRON DI CHLORIDE IRON TETRA    JACSAT-0094-0309
EHYDE RESINS.=   FURFUR AMIDE, A MODIFIER OF UREA - FORM ALD KHMTA6-71-06-021
RIDES IN DI METHYL FORM AMIDE, NITRO METHANE, AND BUTANOL.=  ZOKHA4-0041-2612
ARATION OF N,N-DI ALKYL AMIDES AND CYCLIC IMIDES OF CARBOXYL ZPKHAB-0044-2682
E POL+SYNTHETIC PEPTIDE AMIDES BY DIRECT AMMONOLYSIS FROM TH TELEAY-1972-0101
OF SOME A,B-UNSATURATED AMIDES DERIVED FROM PHENYL ACET AMID RRCHAX-0016-1925
ITIES OF SOME SULF ANIL AMIDES FOR MICE AND RATS.=    +TOXIC ABMGAJ-0027-0141

HESES OF PYRIDYL METHYL AMIDES OF ACETYL TROPIC ACID.=  SYNT APPHAX-0028-0589
G THE BINDING OF SULFON AMIDES TO HUMAN ALBUMIN IN VITRO.=+N APTSAI-0029-0048
HYL BI PHENYL-2-CARB OX AMIDES WITH PER SULFATE.=+SOME N-MET JCPRB4-1972-0118
S SY+PIPERIDINE CARB OX AMIDES WITH POTENTIAL CENTRAL NERVOU JPMSAE-0061-0127
 ALKALINE HYDROLYSIS OF AMIDES.  ALKALINE HYDROLYSIS OF SOME RRCHAX-0016-1925
-ALKYL SU+HYDROLYSIS OF AMIDES.  DILUTE ACID HYDROLYSIS OF N AJCHAS-0025-0303
OTEIN BINDING OF SULFON AMIDES.  EFFECT OF DISEASE AND DRUGS APTSAI-0029-0134
US THIAZINES, AND THEIR AMIDES.  MECHANISM OF MESYL MIGRATIO TETRAB-0028-0167
CTIONS OF 2-CHLORO ACET AMIDES.  SYNTHESIS OF ANILIDES OF    ZORKAE-0007-2572
 POLYMORPHISM IN SULFON AMIDES.=                             JPMSAE-0061-0026
OXYYGEN ETCHING OF POLY AMIDES.=                          OX MKPLA6-0007-1096A
UBSTITUTED O-NITRO BENZ AMIDES.=           +OF SOME N,N-DIS JCPRB4-1972-0097
S OF TERT-ALKANE SULFON AMIDES.=          CONVENIENT SYNTHESI TELEAY-1972-0181
DE OF ACTION FOR SULFON AMIDES.=         NEW ANTI-BACTERIAL MO JPMSAE-0061-0142
RAL SYNTHESIS OF SULFEN AMIDES.=    +SULFUR CHEMISTRY.  GENE TELEAY-1971-4953
 AMINO METHYL THIO FORM AMIDES.=     MANNICH BASES.  TERTIARY TETRAB-0028-0199
ITUTED O-PHEN ACYL BENZ AMIDES.=     STRUCTURE OF N-MONOSUBST LZAKAM-1971-0741
OXIMATES WITH SULF ANIL AMIDES.=   +OF COBALT DI PHENYL GLY  ZOKHA4-0041-2728
ION OF SUBSTITUTED FORM AMIDES.=  +ACIDS BY THE TRANS AMIDAT ZPKHAB-0044-2682
IES OF SUBSTITUTED CINN AMIDES.=+AND ANTI-CONVULSANT PROPERT JPMSAE-0061-0078
-ALKYL SUBSTITUTED ACET AMIDES.=+DILUTE ACID HYDROLYSIS OF N AJCHAS-0025-0303
,3,6-TETRAHYDRO NICOTIN AMIDES.=+5-PHENYL- AND 5-INDOLYL-1,2 CHDCAQ-0273-1776
TIVES OF PHOSPHORYLATED AMIDINES.=      ETHYLEN IMINO DERIVA ZOKHA4-0041-2671
,2- AND 1,3-ALKYLENE DI AMIDO AND DI THIO PHOSPHORUS ACID    ZOKHA4-0041-2808
OSPHITE CONVERSION INTO AMIDO PHOSPHATE.=        +FOR AMIDO PH VMUKA5-0012-0750
```

FIGURE 14: *An example of* KWIC *indexing taken from* Chemical titles.

by Cybernet Timesharing Ltd. Typical search costs are between 50p and £3.00. The whole of the INSPEC data base from 1969 onwards is available on line from Lockheed Information Services DIALOG Retrieval Systems and this can be accessed by direct line or through the Tymshare Network.

The computer allows the publishers of abstracting publications to manipulate the various bibliographical details contained in the references on their data bases in a variety of contexts. Once the necessary software has been developed keyword indexes, using either the natural language of the title or descriptors or keywords assigned by indexers, can be automatically generated with little more than clerical effort. This can equip the abstracting journal with a subject index in each issue, a luxury that was almost unthinkable prior to computer processing. These indexes usually adopt the KWIC (keyword-in-context) or the KWOC (keyword-out-of-context) formats. KWIC is an alphabetical, one item per line, subject index which brings each keyword from a title in turn to a centrally placed and alphabetised filing sequence with the rest of the title 'wrapped around' the keyword to indicate the context in which it is used. The advantage of printing the keyword in context is that the words preceding and following it can be seen as subheadings or modifiers. It is claimed that the compactness of the KWIC format more than compensates for the loss in clarity. Non-significant terms are suppressed as keywords by their inclusion in a stop list. The source of the document is indicated by a document code number given immediately after the KWIC details, this can be translated into the full bibliographical citation by reference to the bibliography section of the index where each code number is listed in turn either alphabetically or numerically according to its nature, along with its corresponding citation. The KWOC format is more conventional in that the keyword is taken out of context and alphabetised on the left hand margin as in the normal subject index. Full bibliographical details are then given against each keyword thus obviating the two consultations of the index usually necessary with the KWIC format. Author indexes can also be easily generated by the computer to provide each issue of an abstracting publication with an author index. *Applied mechanics reviews* have integrated author entries with a KWOC index to produce a single author/subject alphabetical index intended for retrospective search which they dubbed WADEX (word and author index).

The provision of semi-annual and annual author and subject indexes, so essential for the efficient retrospective use of abstracting journals, was an arduous and time consuming process when this work was undertaken manually. Long delays in publication were the norm—in the nineteen fifties the annual subject index of *Chemical abstracts* appeared two years after the completion of the volume it complemented. The computer production of this index has now cut back the delay to less than six months and this pattern is common to other publications. Not only is prompt publication of indexes critical to efficient searching in libraries but also the publication of regular cumulations of these indexes. Again the computer has made the integration of existing monthly or annual indexes into quarterly and quinquennial sequences more feasible than hitherto when the extended manual labour of collation and interfilling rendered cumulations a rarity.

REFERENCES

1 Martyn, J & Slater, P 'Tests on abstracts journals' *Journal of documentation,* 20 (4), 1964 pp 212-35.

2 Martyn, J 'Tests on abstracts journals' *Journal of documentation,* 23 (1), 1967 pp 45-70.

3 Wood, J L et al 'Overlap in the lists of journals monitored by BIOSIS, CAS, and EI' *Journal of the American society for information science,* 23 (1), Jan/Feb 1972 pp 36-8.

4 Wood, J L et al 'Overlap among the journal articles selected for coverage by BIOSIS, CAS, and EI' *Journal of the American Society for information science,* 24 (1), Jan/Feb 1973 pp 25-8.

5 Adams, S & Baker, D B 'Mission and discipline orienting in scientific abstracting and indexing services' *Library trends,* 16 (3), Jan 1968 pp 307-22.

6 Anderla, G *Information in 1985 a forecasting study of information needs and resources* Paris, OECD, 1973 p. 21.

8
Characteristics of the literature: growth, obsolescence, citation patterns and scattering

OVER THE LAST three centuries there has been a dramatic increase in the number of scientific and technical journals. This inflation, which is illustrated graphically in figure 15, has been bemoaned by

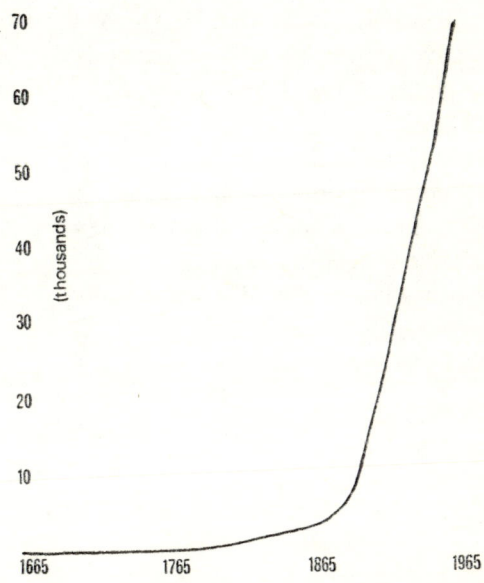

FIGURE 15: *Graph of cumulative numbers of journals published from 1665 to 1970.*

scientists and librarians alike. Librarians are confronted by an ever increasing mass of journal literature which must be acquired and organised to meet the needs of scientists who must spend more and more of their time selecting the pieces of the literature which they must digest to keep themselves at the forefront of their subjects. It has been seen in an earlier chapter that the task of controlling and organising the journal literature has, to the great relief of those involved in information handling, been rendered manageable by the application of mechanised processing. It must be emphasised that the graph represents all the titles which have ever been published, including those which have perished—not just those which are being published currently. The basis for the figures plotted are:

1 1700 30 journals Porter, J R 'The scientific journal—300th anniversary' *Bacteriological reviews*, 28 (3), Sept 1964 p 217.

2 1730 330 journals Barnes, S 'The scientific journal, 1665-1730' *Scientific monthly*, 38, 1934 p 257.

3 1800 750 journals Garrison, F H 'The medical and scientific periodical in the seventeenth and eighteenth centuries' *Bulletin of the institute of medicine, Johns Hopkins University*, 2 (5), July 1934 pp 285-343.

4 1885 5,100 journals The number listed in Bolton, H C *Catalogue of scientific and technical publications*.

5 1895 8,600 journals Bolton, H C *Catalogue of scientific and technical publications* 2nd ed.

6 1920 25,000 journals Number listed in *World list of scientific periodicals* 1925/7.

7 1930 36,000 journals *World list of scientific periodicals* 2nd ed 1934.

8 1950 50,000 journals *World list of scientific periodicals* 3rd ed 1953.

9 1960 60,000 journals *World list of scientific periodicals* 4th ed 1965.

10 1970 75,000 journals From an examination of annual volumes of *World list of scientific periodicals*.

The last examination of the numbers of *existing* scientific and technical journals was made for the Science and Technology Division of the Library of Congress by Gottschalk and Desmond in 1962.[1] They maintained that the number of titles then extant was 35,000 ± ten percent. Periodical was defined as 'a publication intended to be continued indefinitely' but omitted such categories as house journals, report literature, proceedings of international organisations and cover to cover translations. The information used in the census was obtained from current national serial listings published by the various countries throughout the world. Gottschalk and Desmond identified the main publishing countries of scientific and technical periodicals as:

1. United States   6,200 titles
2. Germany (East and West) 3,500 „
3. Japan   2,800 „
4. France   2,800 „
5. USSR   2,200 „
6. United Kingdom   2,200 „

Three years after this census K P Barr[2] of the National Lending Library of Science and Technology (NLL) contested the figure of 35,000 titles. It was suggested that a more realistic figure was 26,000, the total number of journals then held or on order by the NLL, a library which endeavoured to obtain every worthwhile scientific and technical journal and which was successful in meeting ninety five percent of the demands for this literature. Barr agreed that although other titles might be identified in national listings they were not obtainable, usually because publication had either ceased or been suspended. Thus the number of 'live' and useful journals was estimated at 26,000.

Previous to the above estimates of the world's journal population numbers ranging from 50,000 to 100,000 had been suggested. These estimates were extremely misleading as they were usually based on counts of the numbers of titles in the successive editions of the *World list of scientific periodicals* and projections as to how many titles the next edition would contain based on the exponential growth of the journal literature. An important factor was however ignored—the titles identified in the *World list* had a high rate of mortality. Gottschalk reports that thirty three percent of the titles in the third edition were dead, while Barr gives the figure for the fourth edition as forty percent. Some other figures for the mortality rate of journals in particular subject fields are given by Desmond and Gottschalk. For radioactivity literature a death rate of thirty three percent is reported over fifty years while in the field of aeronautics the rate is sixty percent over sixty years.

The rapid inflation of the journal literature is the direct result of the expansion in the numbers of practising scientists working throughout the world and of the increasing pressures exerted on them to publish for prestige and career advancement. Not only are there more scientists active at this time than ever before, De Solla Price[3] estimates that there are seven scientists alive for every eight that have ever been and that the number of existing scientists is doubling every fifteen years, but they have increasingly before them the technological resources necessary for the generation of data—the stuff of the scientific paper. Price[4] asserts that 'the body of world's literature has been growing exponentially for a few centuries and probably will continue at its present rate of about seven percent per annum, there will be about seven new papers each year for every hundred previously published papers in a given field'.

This paper explosion is well evidenced by citing the increase in the number of papers published in the journals of the American Chemical Society over the space of a decade: 4,800 papers in the seven journals being published in 1958 as compared with 7,600 papers in the nineteen journals published in 1968. In the case of new learned society journals the justification for the new journal has invariably been to relieve pressure on an existing publication and to provide outlets for the increasing army of scientists. Gushee[5] has stated that the fundamental set of problems presented to the learned society scientific publishing programmes' is learning when to start the next journal and knowing how to determine which segment of the scientific population is well-funded enough to sustain the new publication so that it will meet the 'general criteria for success that can be articulated from experience while at the same time controlling the size of the parent journal'.

The fragmentation of learned journals into areas of specialisation is classically demonstrated by examining the growth of the American Chemical Society's journal programme. The *Journal of the American chemical society* (*JACS*) was founded in 1879—in 1909 *Journal of industrial and engineering chemistry* (*JIEE*) emerged as an offshoot of the *Journal*. The *News edition* of *JIEE* appeared in 1923 to emerge as *Chemical and engineering news* in

1942, and the *Analytical edition* was founded in 1929 to become *Analytical chemistry* in 1947. Additional offspring of *JIEE* were *Journal of agricultural and food chemistry*, 1953; *Journal of chemical and engineering data*, 1959; the *International edition of industrial and engineering chemistry*, 1959—and finally in 1962, three siblings, *Process design and development, Fundamentals of chemical engineering* and *Product research and development*. The non-industrial children of *JACS* were *Chemical reviews*, 1933, *Journal of organic chemistry*, 1954, *Journal of chemical documentation*, 1961, *Medicinal and pharmaceutical chemistry*, 1962, *Inorganic chemistry*, 1962, *Biochemistry*, 1962, *Macromolecules*, 1968, *and ESCT*, 1968. The *Journal of physical chemistry* which had been founded independently of the society was acquired by ACS in 1933.

In the United Kingdom this fragmentation pattern is also reflected to a lesser extent in the publications of the Chemical Society of London. The *Journal of the Chemical Society* after an undivided existence of a century, split into three divisions in 1966, *Section A: Inorganic, physical and theoretical, B: Physical organic* and *C: Organic*. In 1969 a fourth journal appeared as *Section D: Chemical communications*. Further division took place in 1972 when the four journals were reorganised into six as: *Journal of the Chemical Society (JCS) Dalton transactions* (formerly *A*), *JCS Faraday transactions I & II* (a merger with *Transactions of the Faraday Society*), *JCS Perkin transactions I & II* (formerly *C & B*) and *Chemical communications*. A final and more startling example of this fragmentation from the engineering field is the case of the *Proceedings of the American Society of Civil Engineers* which has split into the Journals of the following Divisions: *Transportation, Engineering, Construction, Engineering mechanics, Hydraulics, Irrigation and drainage, Power, Sanitary engineering, Soil mechanics and foundations, Structural, Surveying and mapping, Urban planning and development, Waterways and harbours, Professional activities*. And in addition many of the Divisions publish their *Newsletters*.

Some estimates of the number of scientific papers published in the world's journal literature have been attempted in recent years.

Vickery[6] in a survey undertaken at the NLL in 1964 estimated that the 26,000 titles then acquired by the library generated 850,000 authored papers. This figure is in step with a contemporary estimate made in 1965 of 950,000 papers,[7] the latter figure however, rather surprisingly, was over a narrower subject field than the NLL survey in that it omitted agriculture and medicine Vickery's figures are based on a sample of 1,118 journals held by the NLL. The concentration of papers in a small percentage of the journals was noted: half of the papers were published in seventy four titles, less than seven percent of the number sampled, and ninety two percent of the papers appeared in 370 of the titles, fifty six percent of the total. Applying these proportions to the 26,000 journals held at the NLL in 1964, less than 15,000 of them would contain authored papers, and 8,000 journals would generate ninety percent of the papers. Thus it was concluded that 'a library with a suitably chosen input at that level should be able to meet ninety percent of the potential demand'. The concentration phenomenon will be examined more fully later in the chapter.

Price's[3] estimate of the number of papers published annually is somewhat lower than the previous figures. He claims that since science began approximately 10,000,000 papers have been published and places the annual output of papers in 1964 at about 600,000, a figure which he maintains is growing by about six or seven percent each year to achieve a doubling every ten years. Thus from the above estimates, placing the annual output of papers in 1964 at 800,000, and accepting Price's doubling rate of ten years, the number of papers published in 1974 was of the order of 1,600,000.

As the journal literature of science and technology has grown exponentially its actual usage and citation patterns have been observed to follow a negative exponential distribution. Papers have a high probability of use and citation in the immediate years after publication but their rate of subsequent use diminishes rapidly as they age. The concept of 'half-life' has been borrowed from the field of nuclear physics to illustrate journal obsolescence. In its original context half-life refers to the 'time required for disintegration of one half the atoms of a sample of radioactive substance'.[8] Its literature connotation refers to the time during which one half of the currently active literature has been published. Science librarians have long been aware of the journal ageing process, studies of obsolescence have appeared in the literature over the last fifty years, one of the first being that made by Gross and Gross who examined the references in a single volume of *Journal of the American chemical society* and reported a halving of the number of references for every fifteen years of increased age.[9] Some years later Fussler, in a study of the characteristics of research literature used by chemists and physicists in the United States, indicated that half of all the references to chemistry were less than eight years old and half the references to physics less than five years.[10] Brown's classic study of scientific serial citation data[11] and Burton's[12] work on engineering serial data present a generalised picture of journal ageing within nine scientific and technological fields. The table below demonstrates the spread of citations in percentages over five decades and gives the estimated half-life of each subject division. The citations used in these studies consist of all the references made in the papers published in the leading primary journals in each field during the years 1953/54.

	Decades 1/2	2/3	3/4	4/5	Half-life
Chemical Engineering	75	88	96		4.8
Mechanical engineering	72	87	94		5.2
Metallurgical engineering	82	93	97		3.9
Mathematics	48	77	89	94	98	10.5
Physics	76	92	98	99	100	4.6
Chemistry	58	77	87	91	95	8.1
Geology	42	68	84	90	94	11.8
Physiology	62	84	94	97	99	7.2
Botany	50	79	90	94	97	10.0

A literature obsolescence graph for the physics figures is given as an example in figure 16.

The obsolescence studies of Brown and Burton were incorporated in the work of Bourne[13] who assembled the results of fifty different studies concerned with the use of literature as a function of age and plotted the graphs of cumulative distribution. The studies used embraced both citation and actual library usage figures. Bourne recognised some of the limitations of citation counting, *eg* the time lag between publication and citation and

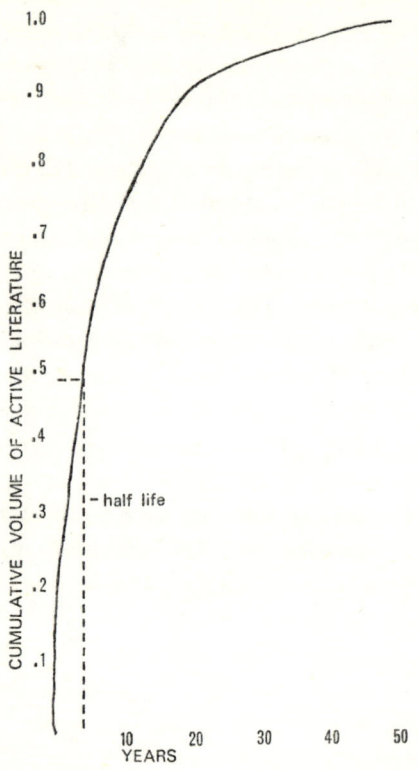

FIGURE 16: *Obsolescence graph for the journal literature of physics.*

the errors arising from using the nominal date of publication rather than the date on which the author actually wrote the paper in which the citation was made, but observed that there was no obvious difference in the curve obtained as a result of applying the two techniques. In view of the severe difficulty involved in obtaining reliable data on actual library usage of a journal, for instance the disturbance of library users and the almost insoluble problems of monitoring in-house browsing and current awareness usage, most studies of journal ageing must rely on citation counting. What Bourne did report was a great variance in the figures that could be quoted as the half-life for a particular field. The curves he plotted represented a smear of possible values for a specific field—the half-lives took on 'a probabilistic rather than a deterministic manner and we now talk of half-lives in terms of variance and best estimates and confidence figures'. The culmination of Bourne's work was to offer what might be considered as 'very fuzzy half-lives rather than easily determined constants'.

The statistical measures that are now being devised for library management applications have been refined considerably since the generalised observations described above were produced. Theoretical models and formulae are being constructed and their possible applications are being discussed in the primary journals covering librarianship and information science. There is however still a gulf between these theoretical approaches and the mainstream of library management practice, this gulf will narrow as more refined theoretical methods are advanced, tested and finally applied in library management decisions.

Brookes has formalised earlier misgivings concerning the absolute half-life of any literature.[14] While agreeing that the ageing process follows a negative exponential distribution he maintains that it is inaccurate to proffer a half-life for the literature of one particular discipline: there will be considerable fluctuations in this 'value' from one library situation to another. No measure of obsolescence can be applied realistically to a particular collection of journals unless it has been generated from an examination of the use of that collection—with all the problems that data collection involves. Generalised measures are useful as guidelines but their application in policy making decisions by individuals as to stock retention or discarding would result in inaccuracies. In any given library situation 'a well randomised sample of at least 580 items is needed to give (and then with a confidence of only ninety five percent) a direct estimate of the half-life which is correct to within ten percent of the true value'.

Librarians will realise that the levels of use of individual runs of journals will vary within a particular subject field: some journals are more useful than others, *Physical review* will produce more papers in a year than for instance *Physica*. Thus a librarian who decides to discard these journals after the same half-life has expired will hardly be making a rational decision—the first title might generate three times as many papers as the second per year and therefore be three times as useful. Brookes has evolved the concept of utility in relation to journal ageing: his 'utility factor' is proportional to the total use that a periodical can expect to attract during the whole of its lifetime in a library collection. The application of this factor seeks to discard tails of journals of equal utility rather than to cut all titles within the subject field after a specified halflife has elapsed.

A further problem of accurately predicting the decline of journal use with age has been demonstrated by Maurice Line[15] who has observed that this is a function of two factors, obsolescence and *growth*. If the number of journals published in a single subject field in 1970 is double the number published in 1960, assuming that the number of papers in each journal remains constant, then the random probability of citation within that field is doubled. Therefore unless the rate of growth of the literature is known, any attempts at predicting a half-life will be further misleading. Line prefers the phrase 'median citation age' to what has been used above for half-life, and substitutes 'corrected half-life' which he defines as 'the halflife as estimated by removing the growth element from the median citation age'. The paper cited above postulates a statistical method of calculating the corrected half-life which, with the foregoing reservations in mind he suggests may be used as 'a background against which library policies can be framed'.

The millions of papers which comprise the literature of science and technology are knitted together into an intricate web of cross citations. Papers beget papers: a scientist will read and assimilate the works of his precursors and peers and use their experience as a catalyst and stimulant to his own activity. Their papers will act as a platform and a springboard for his published work and they will be cited in his papers. His own published works will subsequently be cited by his peers and heirs and so the fabric of the literature will be worked. The exponential growth of the literature described above arises not only from the expanding population of scientists, it can also be explained by the fact that the average paper answers some problems posed by previous papers but it then raises more questions than it has answered.

This interlinkage of papers can now be explored through the *Science citation index* (SCI), a recently developed computer-generated index which complements conventional abstracting and indexing services as it is able to identify relationships between papers that are overlooked by conventional secondary services because of the scattering of papers relevant to any subject field throughout the entire journal literature. An abstracting service covering field A will only be able, for economic reasons, to monitor those journals of immediate relevance and obvious fringe interest to the practitioners in field A—the more remote fringe journals and those from other disciplines which will occasionally publish a paper of interest to workers in field A will not be covered. These are the papers which can be retrieved by using *SCI*, published by the Institute for Scientific Information (ISI). *SCI* which covers over 2,400 journals—the world's primary journals from all scientific and technological fields, generating over 370,000 papers per year—groups papers together according to a reference which they have in common. It is an alphabetical sequence of those articles being currently cited in the world's primary journals arranged under the author cited, each citation is then itself accompanied by a list of the citing papers. *SCI* is published quarterly, cumulating annually, and consists of the *citation index* volumes and the *source index* volumes. The latter provide the full bibliographical references to the papers whose abbreviated citations are listed in the citation index. A five year cumulation of *SCI* for 1965 to 1969 is now available covering the 1,500,000 papers appearing in the source journals for the period which in turn identify 16,000,000 previous citations. The source index itself is a major bibliographical tool in that it is an author bibliography to the papers appearing in the world's leading journals covering science and technology. It is particularly effective for bibliographical checking and

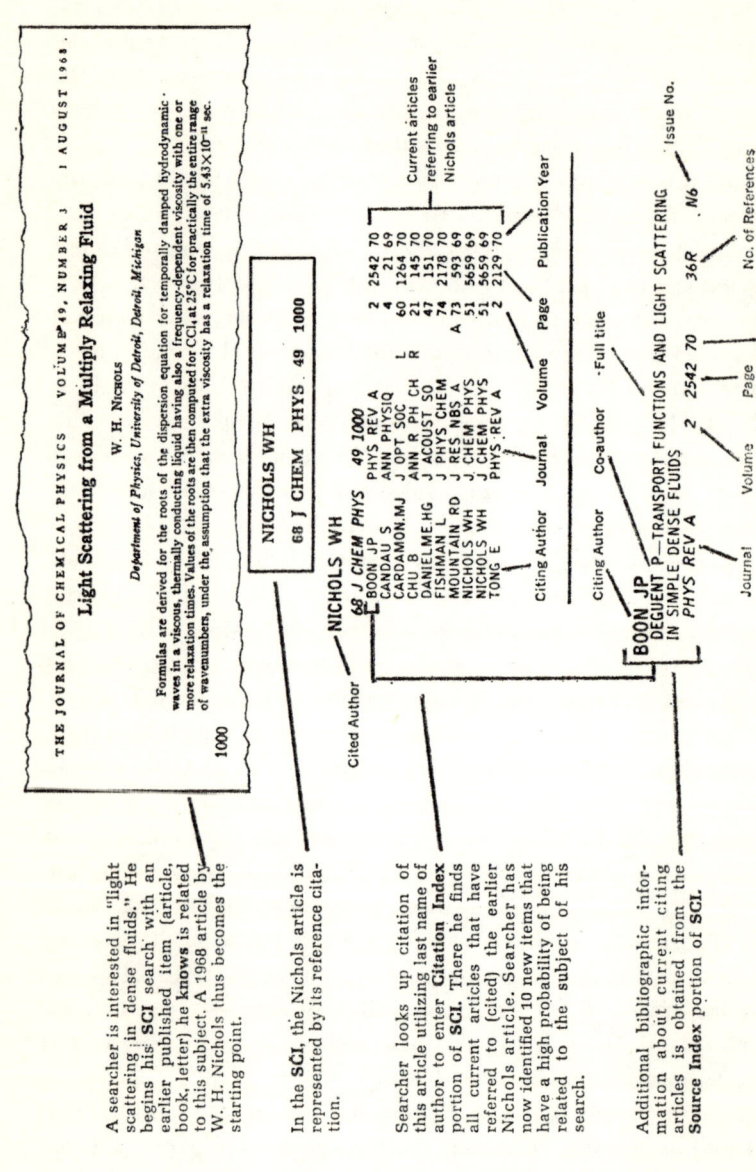

FIGURE 17: *How to use the Science citation index (SCI).*

verification. *SCI* must be approached under the name of an author who is known to have produced a relevant paper in the field under search (figure 17). From this starting point a network of citations emanating from that paper can be traced. Thus the impact of a theory or an idea disclosed in a basic paper—its confirmation, application, elaboration or even its refutation, can be followed. The network of citations is assembled by re-entering the citation index under the name(s) of the author(s) who cited the original paper (paper A in the diagram below) to ascertain whether their works were in turn cited by others—and so on:

PAPER A (original paper) cited by
- B–cited by
  - E–not cited
  - F–cited by
    - I–not cited
    - J–not cited
- C–not cited
- D–cited by
  - G–not cited
  - H–cited by   K–cited by
    - L–not cited
    - M–cited by O
    - N–not cited

thus paper A has yielded 14 citations.

SCI falls into accepted patterns of literature usage—scientists frequently use libraries to follow back the references given in a paper which they have found stimulating. *SCI* reverses this process by allowing the searcher to identify those papers which have subsequently cited a paper of importance. The index is particularly useful when searching within interdisciplinary and newly developing fields where terminology has not yet hardened and where the controlled language used by many secondary services is of dubious value. *SCI* can claim to be self-indexing: it is based on the concept that an author automatically classifies his own work when he cites the works of others. There is a conceptual relationship between paper A and the papers its cites. Some of the problems inherent when using secondary services based on standard indexing languages or classification schemes are not experienced with *SCI*—material is not lost because an indexer has failed to provide for a particular relationship.

ISI also publishes the *Permuterm subject index*, which is based on the natural language titles of the papers from the 2,400 source

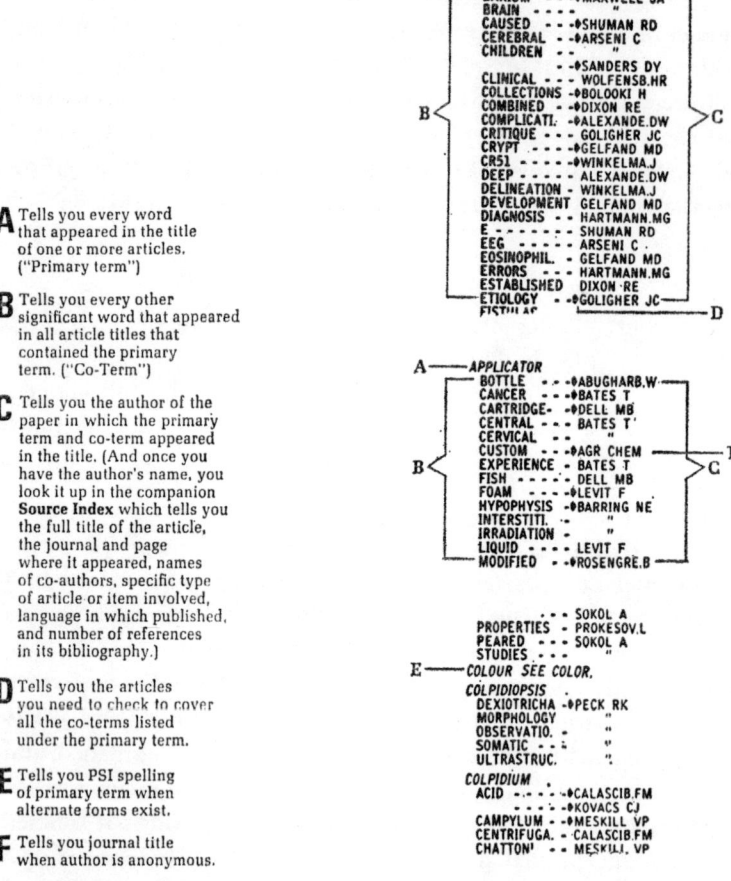

FIGURE 18: *How to use the* Permuterm index (SCI).

journals of *SCI* (figure 18). This index displays the relationships between each significant word used in a title and the other words in the same title in which it appeared—all possible combinations are given. Thus a six word title will generate thirty separate indexing entries for one paper alone. The *SCI* data bases are the most comprehensive multidisciplinary sources now available. They can be obtained on magnetic tape for in-house retrospective search and current awareness systems or they can be accessed through subscriptions by individual profile to the ASCA (Automatic Subject Citation Alert) Service. The ASCA subscriber will produce a profile of his interests consisting of authors' names and keywords; the weekly report he receives from ISI will list the papers published by the authors whose names are in the profile, the papers which have cited those authors, and papers containing the profile keywords in their titles. The last four weeks of the data base can now be accessed on line through the SCISEARCH service via Cybernet Timesharing Ltd. This base is updated weekly making the 30,000 most recently published articles available for search almost immediately after their publication. ISI claim that for most monthly periodicals the bibliographical references will be in the data base before the journal itself is available in libraries. An obvious application of the SCISEARCH is to update an existing bibliography.

The citation patterns of scientific journals have been used to investigate the structure of the literature. An established journal is not a static and isolated phenomenon: it is an active carrier of information within the community of scientific workers. Kessler[16] has demonstrated that a well edited learned journal will belong to a family of journals through the citations given in its papers and it will also in turn serve as a source of citations for other journals within the family. This interflow of information between any two journals is a measure of their subject correlation and the location of a journal within a particular family is a measure of the probability that it will carry a particular type of information. Narin[17] and Carpenter carried out a series of citation examinations of the journal literature in an endeavour to develop empirical models and measures of scientific interrelationships across a range of scientific disciplines. Their source data was derived from the *Journals citation index (JCI)*, a file sorted from *SCI* consisting of a journal by journal tabulation of citings to and from each journal processed for *SCI*. One step citation maps (figure 19) connecting each journal within a discipline to the journal it cites most

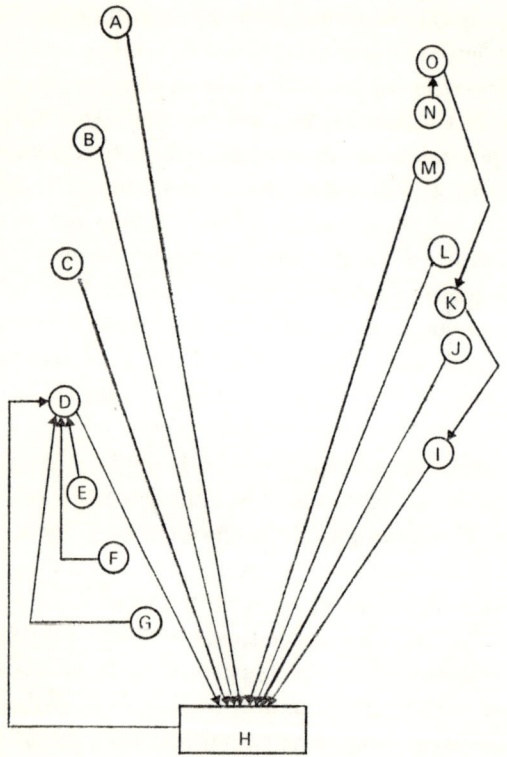

FIGURE 19: *A one-step journal citation map.*

frequently, other than itself, were constructed for a number of disciplines and in each case the key journals became immediately apparent through the number of 'arrows' they attracted. The family tree of the journals for a particular discipline were thus clearly depicted. Two step citation models (figure 20) indicating the first and second most cited journals were next constructed and these made an even clearer exposure of the key journals.

Cross-disciplinary models demonstrated flow of references from one discipline into another and revealed an overall relational sequence thus:

BIOLOGY – BIOCHEMISTRY – CHEMISTRY – PHYSICS – MATHEMATICS AND STATISTICS

with a small number of well known journals as the linkage points, *eg Journal of fluid mechanics* between mathematics/statistics and physics, *Journal of chemical physics* between physics and chemistry and *Biochemistry* between chemistry and biochemistry. This the authors claim is a step towards the establishment of a mosaic of scientific knowledge. A cluster analysis of journal citations again based on *JCI* data partitioned 288 primary journals within

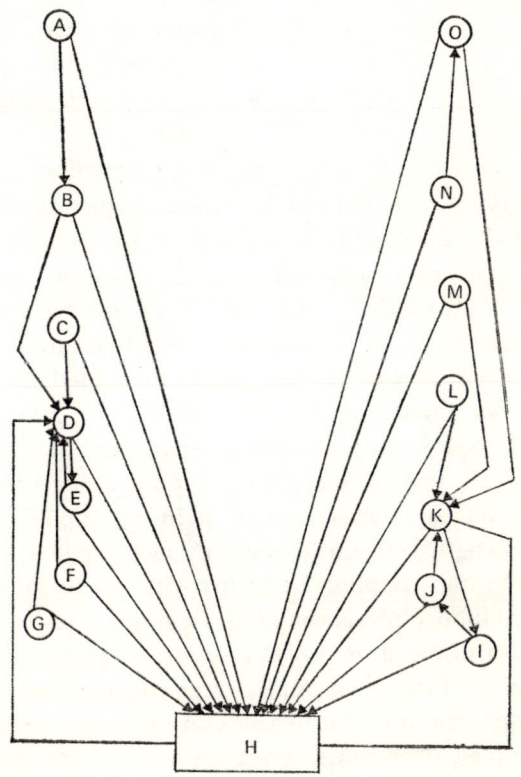

FIGURE 20: *A two-step journal citation map.*

physics, chemistry and molecular biology into a subject classification more precise than that of the discipline thus further delineating the previous mosaic.[18] The journals were so reasonably clustered into subject groupings that, based on the journal titles alone, it was simple to attach a label to each cluster. For the eighty one journals analysed, thirteen clusters were identified, two examples being the acoustics and geophysics and space science groups:

Acoustics	Geophysics and Space
Acustica	*Annales de geophysique*
Journal of the Acoustical Society	*Journal of atmospheric and terrestial physics*
Journal of sound vibration	
Soviet physics acoustics, USSR	*Journal of geophysical research*
	Naturwissenschaften
	Planetary and space science
	Pure and applied geophysics
	Reviews of geophysics and space physics
	Space science reviews

The clusters however did not in all cases reflect subdisciplines, some were characterised by nationality as, for instance, a strong Soviet physics cluster and a strong German chemistry group were identified. These studies are not only of interest to scientific epistomologists, they are of practical value to librarians in that they demonstrate which journals are closely associated by the practising scientific population.

Scientific papers can be related to each other in a number of ways. Kessler[19] minted the phrase 'bibliographic coupling' to denote the sharing of a given paper 'p' by a group of papers, the coupling strength between 'p' and any member of the group is measured by the coupling units (n) between them. A second criteria for bibliographical coupling is a group of papers in which each member of the group has at least one coupling unit to every other member of the group. Bibliographical coupling, which is a fixed relationship in that it depends on the references contained in the coupled documents, has been used as an information retrieval technique to assemble groups of papers on a specific topic.[16] Kessler examined all the papers from a single volume of *Physical review* containing 265 papers and produced 265 related groups by taking each paper in turn as a test case. The number of papers in the groups varied from none to twenty seven with the number of coupling units varying from one in the weakest bond to eleven in the strongest. The practical value of coupling as a retrieval tool has not yet been fully assessed but Kessler claims several advantageous properties for the technique. It is independent of words and language and thus avoids problems arising from terminology and syntax; no expert judgment is required in assembling a group of papers, and the technique does not produce a static classification for a given paper as the groups will undergo changes that reflect the interests of the scientific community.

Small[20] has identified a 'co-citation' principle which is the frequency with which two items of the prior art are cited together by the later literature. Co-citation can be used as a measure of the degree of relationship between papers as observed by practising scientists. The pattern of co-citation may fluctuate just as the intellectual patterns within the field changes. Small maintains that co-citation which links cited documents is analogous to a measure of word association. If frequently cited papers can be taken to represent key concepts inside a discipline, the patterns of co-citation can be used to illustrate in detail the relationships existing between those concepts and through the changing pattern of co-citation it might be possible to monitor the development of scientific fields. Thus when viewed retrospectively, co-citation 'could provide clues to understanding the mechanism of speciality development'. A possible application of the idea in information retrieval might be that co-citations, identifiable by using *SCI*, could be used to isolate a cluster of core articles for a particular topic which could then serve as a profile for that topic in an SDI system.

Figure 21 represents the interconnections between a set of papers on a particular topic—such a network can easily be built up by using the cycling technique in *SCI*. The diagram demonstrates that papers 1, 2 and 5 have attracted much attention and obviously their authors have had a considerable impact on other workers

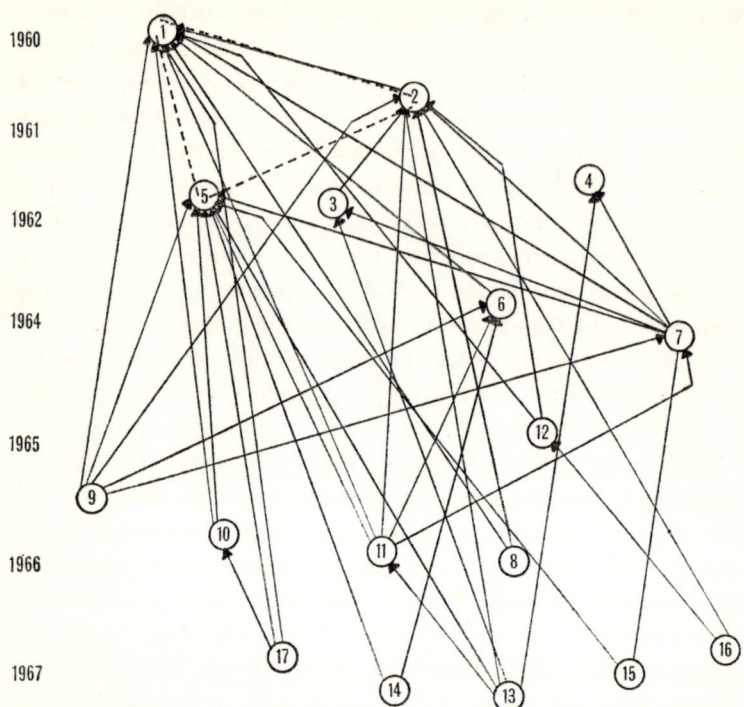

FIGURE 21: *A paper citation network.*

in the field. There is a probability that papers 9 and 11 are closely related as there is strong bibliographical coupling—papers 1, 2, 5, 6 and 7 are cited by both. Co-citation is seen in papers 1 and 5 which are co-cited by 7, 9, 10, 11 and 17, between 1 and 2, co-cited by 7, 8, 9, 11 and 12, and 2 and 5, by 7, 9, 11 and 13.

The papers relevant to any given subject are, apart from a tight concentration of papers in a small number of productive 'core' or 'nuclear' journals, scattered widely over the whole range of scientific and technical journals. This phenomenon of *scattering* was first investigated by S C Bradford, a former librarian of the Science Museum Library.[21] Bradford analysed the sources of references contained in the current bibliographical aids covering applied geology and geophysics and lubrication over a number of years and concluded that 'the articles of interest to any specialist must occur not only in the periodicals specialising in his subject, but also, from time to time, in other periodicals which grow in number as the relation of their fields to his lessens and the number of articles in each decreases. He observed that the journals could be grouped into three zones with each zone contributing about the same number of papers. For applied geophysics

zone 1 (nuclear)	9 journals produced	429 papers
zone 2	59 ,,	,, 499 ,,
zone 3	258 ,,	,, 404 ,,

for lubrication

zone 1 (nuclear)	8 journals produced	116 papers
zone 2	29 ,,	,, 133 ,,
zone 3	127 ,,	,, 152 ,,

The graph of this distribution was plotted on semi-logarithm paper with the journals in decreasing order of productivity on the abscissa and the cumulative total of papers contributed by the journals on the ordinate and after an initial rising curve the graph was remarkably close to a straight line. From the graph Bradford observed that apart from those produced by the first group of journals the aggregate of papers on a given subject is proportional to the logarithm of the number of producers. On the foregoing evidence he postulated his law of scattering which states that if a large collection of papers on any given topic is ranked in decreasing order of productivity of papers three zones can be marked off with each containing approximately one third of the total number of papers on the subject. The first, or nuclear zone, p, will contain a small number of highly productive journals, the second, p_1, a larger number of moderately productive journals, and the third, p_2, a much larger number of journals of low productivity. The ratio of $p:p_1:p_2$ is as $1:a:a^2$ where 'a' approximates fairly closely to the number 5.

Bradford maintained that 'the whole range of periodicals acts as a family of diminishing kinship, each generation being greater in number than the preceding, and each constituent of a generation producing inversely according to its degree of remoteness'. He modestly observed that his law conformed to the mathematicians' criterion of being 'of no possible value whatsoever'—but he did note that the scattering phenomenon explains why no abstracting journal attempting to cover a discipline can realistically achieve 100 percent coverage just as no special library can ever meet the potential needs of all of its clients' demands without becoming a general collection of scientific journals.

Bradford's law has attracted much recent attention. It was initially refined by Vickery[22] who demonstrated that it could apply not only to three zones, but, with suitably modified values of 'a' to four, five, or any number of zones. Latterly information scientists have sought to uncover its underlying formula and to apply this to the management of collections of scientific journals. This developmental work has been carried out primarily by Leimkuhler[23] and Brookes.[24, 25] Leimkuhler derived a model from the law which he termed the Bradford distribution and used this to predict the reference yield of abstracting services in terms of papers and their sources in the field of thermophysical property data. His expression of the distribution was

$$F(x) = \frac{ln(1 + \beta x)}{ln(1 + \beta)}$$

where

x = the fraction of the document collection

$F(x)$ = the proportion of total productivity contained in the fraction 'x'

β = a constant related to the document collection.

Leimkuhler's analysis of Bradford's law offered a formula which could be used to express the distribution of papers relevant to a topic over the range of journals known to be productive for the topic. Leimkuhler also noted that Bradford's law and Zipf's law,

$$fr = C$$

which relates to word frequency in linguistic samples, where 'f' is the frequency of occurrence of the word ranked 'r', and 'C' is a constant for the sample, were essentially the same thing.

Brookes, while noting previous work on scattering, further analysed both the Bradford and the Zipf laws and from these sources formulated his combined Bradford-Zipf distribution. He identified and described the separate components of the joint distribution both mathematically and verbally. The Zipf situation is the linearity B/C of graph (figure 22) in which there are no restraints on the productivity of contributing sources. The

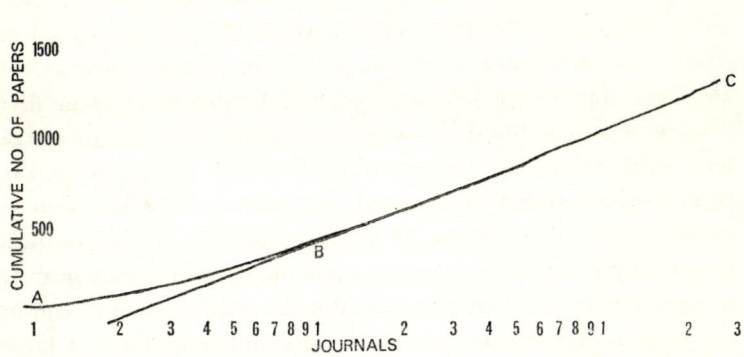

FIGURE 22: *Graph of the cumulative distribution of papers over journals.*

Bradford initial rising curve A/B on the graph assumes the Zipf linearity outside the nuclear zone. This he explains by observing that when the first papers on a new subject are written they are offered to a limited number of appropriate journals and accepted. These papers attract further related papers but soon other papers, perhaps on applications of principles within the subject field, appear in journals of marginal interest and related subject fields. However, by this time the original journals have become established and assumed the role of core journals for the field. Thus the constrictions operating within the nuclear zone are seen. The Bradford-Zipf distribution is given as

$$R(N) = \frac{N \log xn}{s}$$

where $R(N)$ is the total number of papers covering a field and N is the rank of the last journal; 's' is the slope of the linearity CB (see figure 22).

Brookes[25] has foreseen several practical applications for his distribution, the most potentially interesting of which to the library manager is that of being able to assess the coverage of a bibliography or an abstracting service—a check which has hitherto not been available.

Cores or nuclei of journals are as Bradford observed ubiquitous, they occur in all subject disciplines. Two examples of the coverage of abstracting services over two of the major scientific fields will serve to illustrate this fact. An analysis of the papers covered in *Physics abstracts*[26] indicated that 405 journals contributed 20,287 abstracts, but 10,573 of the abstracts, slightly more than fifty percent, were produced by twenty titles, only five percent of the total, and seventy five percent of the coverage was given by fifty titles, twelve percent of the total. A similar pattern was seen to obtain in chemistry where about eight percent of the journals covered by *Chemical abstracts*[27] contributed more than seventy five percent of the total abstracts. But the overall pattern seen in the totality of journals covering science and technology at large is also one of concentration into a highly fertile core. This is admirably illustrated by examining the journals processed by SCI —when the source journals are ranked in decreasing order of the times each is cited, the top twenty five titles, one percent of those covered, yield twenty four percent of all citations, the top 152 titles account for fifty percent of the citations, 767 journals provide seventy five percent of the citations and 2,000 titles eighty five percent.[28] Martyn and Gilchrist's[29] analysis of citations to British scientific and technical journals for 1963/4 which also used SCI data demonstrates an even greater concentration into a core. The top twenty five titles generated sixty five percent of all the citations, fifty titles provide seventy seven percent of the citations, and 163 titles, or only nine percent of all British scientific and technical journals, provided ninety five percent of the total citations.

A similar pattern of concentration emerges when the literature is analysed by a different method. Garfield[28] has shown that twenty journals provided twenty percent of all the papers published in the 2,200 titles covered by SCI in 1969, 100 titles provided forty three percent of the papers and 500 titles seventy percent.

Garfield's work of the SCI data led him to formulate a 'law of concentration' which generalises Bradford's law to the interdisciplinary situation.[30] He maintains that about 1,000 journals will contain the leading titles in all specialisations and will, in addition, contain a large proportion of the papers published within a particular field. Thus, as Vickery observed earlier, a good multidisciplinary collection of journals which will cover the needs of scientists need contain no more than a few hundred titles. Garfield states that a basic list of 500 to 1,000 titles will account for eighty percent to 100 percent of all journal references. The law of concentration, therefore, postulates for the whole of science what Bradford's law postulates for a single discipline. Although this in one sense corroborates Bradford, it does in some measure counteract the inconvenience caused to the users of the literature by 'scatter'. Some 50,000 scientific and technical periodicals are at present being published throughout the world but it can be seen from Garfield's work and from the analyses that have been made of the productivity of journals covered by abstracting services that the vast majority of these journals produce only a small percentage of the papers documenting original scientific advances. They can be considered as little more than background noise in terms of productivity. The cores of journals identified by citation analyses are, of course, the primary journals, those which have been used by the practising scientific population for the original information which they publish. The journals standing outside the cores in all disciplines and also out of the interdisciplinary core cannot, however, be dismissed as of no value. Although they may not publish original authored papers they are useful in that they provide news and summarised accounts of the chief events and advances inside and related to their field of coverage. They are the newspapers of science and technology, they may be heavily used but they will be rarely cited in other journals. The highly popular *New scientist*, whose circulation is in excess of 50,000, for instance ranks well below the top 1,000 journals of SCI. Like others of its genre covering either science and technology in general or a specific area of knowledge, it is produced not as an organ of original communication but as a medium for disseminating news information.

REFERENCES

1 Gottschalk, C M and Desmond, W F 'Worldwide census of science and technology serials' *American documentation,* 14 (3), 1963, pp 188-94.

2 Barr, K P 'Estimates of the number of currently available science and technology periodicals' *Journal of documentation,* 23 (2), June 1967, pp 111-6.

3 De Solla Price, D K *Little science, big science* New York, Columbia University Press, 1963.

4 De Solla Price, D K 'Networks of scientific papers' *Science,* 149, 30th July 1965, pp 510-5.

5 Gushee, D E 'Problems of the primary journal' *Journal of chemical documentation,* 10 (1), Feb 1970, pp 30-2.

6 Vickery, B C 'Statistics of scientific and technical articles' *Journal of documentation,* 24 (3), Sept 1968, pp 192-5.

7 Carter, C F et al *National document handling systems for science and technology* New York, Wiley, 1967.

8 *International dictionary of physics and electronics* Princeton, NJ, Van Nostrand, 1956, p 406.

9 Gross, P L K and Gross, E M 'College libraries and chemical education *Science,* 28th Oct 1927, pp 385-9.

10 Fussler, H H 'Characteristics of the research literature used by chemists and physicists in the United States' *Library quarterly,* 19, Jan 1949, pp 19-35.

11 Brown, C H *Scientific serials* (ACRL Monograph no 16). Association of College and Research Libraries, Chicago, Ill, 1956.

12 Burton, R E and Kebler, R K 'The half-life of some scientific and technical literatures' *American documentation,* 11 (1), 1960, pp 18-22.

13 Bourne, C P 'Some user requirements stated quantitatively in terms of the 90% library' (In: Kent, A (Ed) *Electronic information handling systems* Washington, DC, Spartan Books, 1965, pp 93-110).

14 Brookes, B C 'Obsolescence of special library periodicals:

sampling errors and utility contours' *Journal of the American society for information science,* 21 (5), Sept/Oct 1970, pp 320-9.

15 Line, M 'The half-life of periodical literature' *Journal of documentation* 26 (1), Mar 1970, pp 46-54.

16 Kessler, M M 'Some statistical properties of citations in the literature of physics' (In: M E Stevens (Ed) *Statistical association methods in mechanised documentation* Symposium Proceedings, 1964. Washington, DC, National Bureau of Standards, 193-8, 1965).

17 Narin, F C et al 'Interrelationships of scientific journals' *Journal of the American society for information science,* Sept/Oct 1972, pp 323-31.

18 Carpenter, M P and Narin, F C 'Clustering of scientific journals' *Journal of the American society for information science* Nov/Dec 1973, pp 425/36.

19 Kessler, M M 'Bibliographic coupling between scientific papers' *American documentation* Jan 1963, pp 10-25.

20 Small, H 'Co-citation in the scientific literature' *Journal of the American society for information science* July/Aug 1973, pp 265-69.

21 Bradford, S C *Documentation* London, Crosby Lockwood, 1948.

22 Vickery, B C 'Bradford's law of scattering' *Journal of documentation,* 4 (3), 1948, pp 198-203.

23 Leimkuhler, F F 'The Bradford distribution' *Journal of documentation,* 23 (3), Sept 1967, pp 197-207.

24 Brookes, B C 'The derivation and application of the Bradford-Zipf distribution' *Journal of documentation,* 24 (4), Dec 1968, pp 247-65.

25 Brookes, B C 'Bradford's law and the bibliography of science' *Nature,* 244, 6th Dec 1969, pp 953-6.

26 Keenan, S and Atherton, P *The journal literature of physics* New York, American Institute of Physics, 1964.

27 Wood, J L *The parameters of document acquisition at Chemical Abstracts Service* (A paper presented at the American

University 8th Annual Institute of Information Storage and Retrieval, Washington, DC, 14th-17th Feb 1966.)

28 Garfield, E 'Citation analysis as a tool in journal evaluation' *Science*, 178 (4060), 3rd Nov 1972, pp 471-9.

29 Martyn, J and Gilchrist, A *An evaluation of British scientific journals* London, Aslib, 1968.

30 Garfield, E *Current contents*, No 31, 4th Aug 1971, pp 5-6.

Index

Abbreviations for journal titles 74-7
Abstracting services:
  back-up services to 89-90
  computer-aided publication of 84-100
  coverage of 86-8
  currency of 84
  current-awareness use of 84, 93-6
  development of 81-3
  discipline-oriented 90-1
  indexes to 100
  indexing of 90-100
  mission-oriented 90-1
  numbers of 83
  organisation of 90-100
  overlapping of 86-88
  problems of 86-90
  retrospective searching of 78, 96
Abstracting services 83
Abstracts:
  author 72
  auto- 73
  defined 68-9, 72-3
  features of 68
  indicative 68-70
  informative 68-9, 71
  telegraphic 72
  use of 77-80
Academie des Sciences 17
Acta diurna 11
Acta eruditorum 16
Acta medica et philosophia hafniensia 17
Adams, S 90, 100
Advancement of science 24
Aeronautical engineering: lists of periodicals 61
Agriculture: lists of periodicals 61
Albritton, E C 47, 48
Allgemeines journal der chemie 20
American Chemical Society:
  journals 104-5
  microfilm editions of 45

American Mathematical Society: Mathematical Offprint Service (MOS) 47
American mechanic's magazine 23
American Society of Civil Engineers: journals 105
American Society for Testing and Materials 76
Analytical review 82
Anderla, G 100
Analen der chemie 21
Annales de chemie 20
Annales de chemie et de physique 20
Annales de physique 20
Annuaire de la presse française 56
Anthropology: lists of periodicals 62
Applied mechanics reviews 99-100
Applied science and technology index 85
Archiv fur die theoretische chemie 20
ASCA (*Automatic Subject Citation Alert*) 116
Aslib 86
Astronomoscher jahresbuch 85
Atherton, P 127
Author abstracts 72
Auto-abstracts 73
Auxiliary publication schemes 45
Ayer, N W & Sons directory: newspapers and periodicals 55

Bacon, F 12
Baker, D B 90, 100
Baker, E A 64
Barries, S 102
Barr, K P 103, 126
Bernal, J D 46, 51
Bertholm, T 17
Betik, A L 80
Bibliographic coupling 118-9
Bibliographical citations:
  abstracts 68
  citation maps 116-8

Bibliographical citations: *(contd.)*
  co-citations 119
  elements of 73-8
  patterns of 106-25
  review papers 35
Bibliographical guide to refrigeration 66
Bibliographical scattering 120-5
  as a constraint to information gathering 42, 78
Bibliographies of scientific journals 52-67
  early journals 58
  house journals 67
  national lists 53-7
  national union catalogues 52
  subject guide 58
  subject lists 60-7
Bibliography of food 64
Biological abstracts 69, 85, 87, 92-3, 95
BioSciences Information Service 87-8, 91, 94-5
Bolach, D H 61
Bolton, H C 58, 102
Botanical magazine 21
Botany: lists of periodicals 62
Bourne, C P 107-9, 126
Bradford, S C 120-3
Bradford-Zipf distribution 122-4
Bradford's law 120-5
British abstracts 82
British Association for the Advancement of Science 24
British Association of Industrial Editors yearbook 67
British Library Lending Division:
  loan and photocopying services 89
  research newsletter collection 51
  supplementary publication scheme 45
British periodicals of medicine 65
British technology index 85
British union catalogue of periodicals (BUCOP) 52, 60
Brookes, B C 109-110, 122-4, 126
Brown, C H 60, 107, 126
Bulletin signaletique 69, 85
Burton, R E 107, 126

CA condensates 85, 94, 96
Campbell, T H 42, 51
Canada: listings of scientific journals 55

Carpenter, M P 115, 127
Catalogue des publications françaises: scientifiques, techniques 56
Catalogue of scientific and technical periodicals, 1665-1895 58
Catalogue of scientific serials of all the natural, physical and mathematical sciences, 1633-1876 58
Catalogue of scientific journals 52-67
Chemical abstracts:
  computer-aided production of 91-2, 100
  coverage of 87-9, 124
  expansion of 92
  founded 83
  informative abstracts 69
  subject groupings 78
  use of coden 76
Chemical abstracts service source index 62
Chemical-biological activities 91, 94
Chemical Society 21, 46
  journals 105
Chemical titles 91, 96, 98
Chemisches journal fur die freund der naturlehre 20, 82
Chemisches zentralblatt 82
The chemist 20
Chemistry: lists of periodicals 62
China: listings of scientific journals 55
Chinese scientific and technical serial publications in the Library of Congress 56
Classified directory of Japanese periodicals: engineering and industrial chemistry 57
Citations, bibliographical:
  abstracts 68
  citation maps 116-8
  co-citations 119-20
  elements of 73-8
  patterns of 106-25
  review papers 35
Co-citations 119-20
Coden 75-6
Collets Holdings. Gazety i zhurnaly SSSR 57
Collison, R L 67, 68, 80
Commercially published journals:
  controlled-circulation 38-9
  primary 35-6
  technical and trade 36-8

Communications journals 33-4
Computer-aided publication of abstract services 84-5, 90-100
Computer-based information services:
 availability of data-bases 96-7
 current awareness 93-6
 on-line access 97-8
 retrospective searching 96
Computers: lists of periodicals 63
Controlled-circulation journals 38-9
Corante 11
'Core' journals 42, 78
Council of Biology Editors 32
Crell, L von 20, 82
Crell's chemical journal 83
Critical scientific journalism 17
Current agricultural serials 61
Current awareness aids 85, 93-6
Current contents 85
Current serials received by the NLL 59
Curtis's botanical magazine 21
Cybernet Timesharing Ltd 99, 115

Data-bases, computer produced 91, 93, 96-7
Delays in publication of scientific papers 42-3
De Sallo, M 12-4
Desmond, W F 102, 126
De Solla Price, D 35, 41, 104, 106
Deutsche presse 56
Directory of Canadian scientific and technical periodicals 55
Directory of Japanese scientific periodicals 57
Discipline-oriented abstracting services 90-1

Editorial boards 33, 36, 43
Electrical engineering: lists of periodicals 63
Electrical engineering abstracts 91
Electronics: lists of periodicals 63
Elsdon-Drew, R 43, 51
Endeavour 40
Engineer 24
Engineering 24
Engineering index 63, 69, 83, 87, 92-3
Engineering: lists of periodicals 63
Entomology: lists of periodicals 63
Evaluation of British journals 54

Faxon's librarians guide to periodicals and American subscription catalog 55
Fechner, G T 82
Finer, R 96
Food technology: lists of periodicals 64
Die fortschritte der physik 82
Foskett, D J 64
Fowler, M 58
France: listings of scientific journals 56
Frankfurter journal 11
Fussler, H H 107, 126

Garfield, E 124-5, 127, 128
Garrison, F H 11, 18, 19, 58, 102
Gazety i zhurnaly SSSR (Collets Holdings) 57
Gebbie house magazine directory 67
General purpose journals, scientific societies 34
Geology: list of periodicals 64
Geoserials 64
German scientific periodical literature 19
Germany: listings of scientific journals 56-7
Gilchrist, A 54, 124, 128
Giornale de litterati d'Italia 17
Gottingische zeitung von gelehrten sachen 18
Gottschalk, C M 102, 126
Graves, C 31
Great Britain: listings of scientific journals 54
Green, D 49-51
Gross, P L and E M 107, 126
Guide to computer-based information services 96
Guide to current British journals 54
Guide to scientific periodicals 58
Guide to special issues and indexes of periodicals 55
Guide to the world's abstracting and indexing services in science and technology 83
Gushee, D E 104, 126

'Half-life' of journal literature 106-110
Harberer, I I 40, 41, 67
Hawelke, A 63

Herlin, J P 42, 51
Herschman, A 44, 51
'Hidden colleges' 12
Hooke, R 16
Houghton, B 64
House journals 39-40
  lists of 67
'House-style' 33

Index medicus 91
Indexing services 85
Indicative abstracts 68-70
  example of 70
Information Exchange Groups (IEGS):
  advantages of 48
  demise of 49-50
  dissemination processes 47-8
  founded 47
  membership of 47
  opposition to 48-9
  scientific journals, attitude to 49-50
Informative abstracts 68-9, 71-2
  example of 71
INSPEC 91, 94-7, 99
Institute for Scientific Information 85, 89
Institution of Civil Engineers 28-30
Institution of Electrical Engineers 44, 83, 91, 95
Institution of Mechanical Engineers 30-1
International Council of Scientific Unions, Abstracting Board 84-5
International Federation for Documentation 60, 83
International Serials Data System (ISDS) 76
International union list of Communist Chinese serials, scientific 55
Irregular serials and annuals 53

Japan: listings of scientific journals 57
Johnson, J 82
Journal des savants 14
Journal des scavans 12-4, 16
Journal der physik 21
Journal fur praktische chemie 20
Journal of the American Chemical Society 104-5
Journal of the Chemical Society 105
Journal of the Franklin Institute 23
Journals citation index 116

Katz, D B 55
Keenan, S 127
Kessler, M M 115, 118, 119, 127
Kuney, J H 41
KWIC index to English language abstract and indexing publications received by the NLL 84
KWIC indexes, 91, 98-9
KWOK indexes 99

Language barriers 69, 79
Lavoisier, A L 20
Lawrence, G H M 62
Learned society journals 32-5, 104-5
Le Fanu, W R 65
Leimkuhler, F F 122, 127
Leitfaden fur press und werbung 56
'Letters' journals 33-4
L'Hermite, R 45, 51
Library of Congress 52-3
Library of Congress, Science and Technology Division 59
Liebig, Justus 21
Line, M 110, 127
Linnean Society 27
List of scientific and technical periodicals received from China 55
List of scientific and technical serials currently received by the Library of Congress 59
List of serials covered by members of the NFSAIS 60
Lockheed Information Services 99
Lockyer, N 24, 26
Luhn, H P 73, 80

Maddox, J 26
Maizell, R E 65
Martyn, J 54, 86, 88, 100, 124, 128
Mason, P C R 57
Mathematical Offprint Service (MOS) 47
Mathematical reviews 47
Mathematics: lists of periodicals 64
Mechanical engineering: lists of periodicals 64
Mechanics' Institutes 23
Mechanic's magazine 22-3
Medicine: lists of periodicals 64-5
MEDLARS 72, 91, 97
MEDLINE 97
Mercurius Gallo-Belgicus 11
Mercury 11

Mezhenko, Y A 57
Microfilm 44-5
Microform, primary publication in 44
Military science: lists of periodicals 65
Mineralogy: lists of periodicals 65
Miscellanea curiosa 16-7
Mission-oriented abstracting services 90-1

Narin, F C 115, 127
National Diet Library, Japan 57
National Federation of Science Abstracting and Indexing Services 60, 83, 84
National Institutes of Health 47, 49-50
National Lending Library for Science and Technology 55, 59, 84, 89
National Library of Medicine 91, 97
National listings of scientific journals 53: Canada 55; China 55-6; France 56; Germany 56; Great Britain 54; Japan 57; United States 55; USSR 57
National Reference Library of Science and Invention 59
National Science Library, Canada 55
National union catalogues 52-3
Natural history: lists of periodicals 65-6
Nature 25-7
  letters to 33
  opposition to Information Exchange Groups (IEGs) 50
Der naturfoscher 18
New periodicals titles 60
New serial titles 53
Newsletters, research 50-1
Newspaper press directory 54
Newspapers 11
Nuclear science: lists of periodicals 66
Nuova raccolta opuscoli scientifici 18

OATS (Original tear sheet service) 89-90
Obsolescence of journal literature 106-110
Oldenburg, H 14
On-line computer-based information retrieval services 97, 99

Peking gazette 11

Periodica chimica 63
Periodical literature of physics 66
Periodical publications in the National Reference Library of Science and Invention 59
Permuterm subject index 113-5
Pflucke, E H M 63
Pharmaceutisches central-blatt 82
Phelps, H R 42, 51
Phillips, J R 31
Philosophical magazine 21
Philosophical transactions of the Royal Society 14-7
Physical review letters 34
Physical Society 46
Physics: lists of periodicals 66
Physics abstracts 83, 85, 91, 214
Physikalische berichte 82, 85
Polymer science and technology 91
Porter, J R 39, 102
Le Pour et le contre 18
'Prestige' journals 39-40
Primary journals 32-6
Primary publication in microform 44
Proceedings of the American Society of Civil Engineers 105
Proceedings of the Institution of Civil Engineers 30

Raccolta d'opuscoli scientifici e fililogici 18
Railway engineering: lists of periodicals 66
Receuil des memoires et conferences sur les arts et les sciences 16
Referatifnyi zhurnal 85
Refereeing of papers 33, 43
Refrigeration engineering: lists of periodicals 66
Research newsletters 50-1
RETROSPEC 97
Review journals 35
Rosenbaum, M 80
Royal Society of London 14-6
Royal Society Scientific Information Conference, 1948 46
Rozier, Abbe 18
Rubber and plastics: lists of periodicals 66
Russian journals of mathematics 64
Russkaya tekhnicheskaya periodica 1800-1916 57

Saarbuch, W E subscription catalogue 56
SATCOM report 35, 41
Scattering, bibliographical 120-5
 as a constraint to information gathering, 42, 78
Scherer, A N 20
Science 33, 50
Science abstracts 83
Science citation index (SCI) 54, 111-9
Science Museum, Great Britain: periodicals on open access 59
Science Reference Library periodical news 59
Scientific American 23
Scientific and technical serial publications of the Soviet Union, 1945-1960 57
Scientific papers:
 citation patterns 106-25
 concentration of 106, 124-5
 delays in publication of 42-3
 growth of numbers 92-3, 104-6, 111
 length of 42
 obsolescence of 106-10
 refereeing of 33
 review 35
 scattering of 42
 separates 46
Scientific periodicals:
 access to 52
 alternatives to 42-5
 bibliographies of 52-67
 commercially published 35-9
 communications journals 33
 controlled-circulation 38-9
 core journals 42, 78, 124-5
 expansion in numbers 18-9, 24, 101-4
 forms of 32-41
 German 19
 house journals 39-40
 Information Exchange Groups and 47-50
 language distribution of 103
 letters journals 33
 mortality rate of 103
 national listings of 51-7
 national union catalogues 52-3
 origins 11
 prestige journals 39-40
 primary journals 32-8
 problems of 42-4
 review journals 35

 roles of 19, 43-4
 specialised 12, 17, 19
 subject guide 53
 subject listings 61-7
 subject rationalisation of 45
 technical and trade 36-8
Scientific serials 60
Scientific and learned societies 11, 17
 journals 32-5
 specialised 27-31
Scisearch 115
Scriveners Company of London 11
Scudder, S H 58
SDI services 93-6
Selective dissemination of information services 93-6
'Separates' (papers) 46
Shih, B P N 55
Siegel, F 66
Slanted abstracts 69
Slater, P 86, 88, 100
Small, H 119, 127
Snyder, R L 55
Societies, scientific 11, 17
 journals 32-5
 specialised 27-31
Special Libraries Association 55
Standard periodicals directory 55
Standards for abbreviations of journals titles 74-7
Steeves, H A 64
Sticker, B 56
Supplementary publication schemes 45

Technical journals for industry 60
Telegraphic abstracts 72
Telford, T 27
Textiles: lists of periodicals 67
Trade and technical journals 36-9
 antecedents 23

UK Chemical Information Service 93-4
UKCIS 93-4
Ulrich's international periodicals directory 53, 84
UNESCO 76
Union list of serials in libraries of the United States and Canada 52-3
UNISIST 76
United States: lists of scientific journals 55
USSR: lists of scientific journals 57

Verzeichnis deutscher wissenschaftlicher zeitschriften 56
Vickery, B C 106, 122, 125, 126

WADEX 100
Watson, D 46
Weinberg report 35, 41
Wildlife diseases 44
Willings press guide 54
Wood, J L 100, 127
Woodward, D 54

Wooster, H 42, 51
World list of scientific periodicals 52, 60, 74-5, 102-3
Wu, J 56
Wyatt, H V 51

Zikeev, N T 57
Zipf's law 122
Zoology: lists of periodicals 67
Zwemer, R L 41

Q
225.5
H57

JAN 17 1977